Pascal's Wager

Pascal's Wager

THE MAN WHO PLAYED
DICE WITH GOD

James A. Connor

HarperSanFrancisco
A Division of HarperCollins*Publishers*

HarperCollins books may be purchased for educational, business, or sales promotional use. For information please write: Special Markets Department, HarperCollins Publishers, 10 East 53rd Street, New York, NY 10022.

HarperCollins Web site: http://www.harpercollins.com

HarperCollins®, 📖 ®, and HarperSanFrancisco™
are trademarks of HarperCollins Publishers.

FIRST EDITION

Library of Congress Cataloging-in-Publication Data is available.

ISBN–10: 0-06-076691-3
ISBN–13: 978-0-06-076691-7

06 07 08 09 10   RRD(H)   10 9 8 7 6 5 4 3 2 1

CONTENTS

	PASCAL	FRANCE	EUROPE AND THE NEW WORLD
1588	Birth of Étienne Pascal, father of Blaise Pascal, in Clermont, in the Auvergne region.	May 12: Day of Barricades in Paris. Duc de Guise seizes the city. July: Henry III capitulates to the duc de Guise.	Spanish Armada nearly succeeds.
1605		Huguenot refugees resettle in Netherlands, Ireland. *Don Quixote* is published.	James River colony founded in Virginia. Czar Boris Godunov dies. Paul V becomes pope.
1616	(?) Étienne Pascal marries Antoinette Begon.		Pocahontas arrives in England. Copernicus's *De revolutionibus* is placed on the Index of Forbidden Books.
1617	Birth of Anthonia Pascal, who dies days after her baptism.		Saint Rose of Lima dies in Peru. Pocahontas dies.
1619	Étienne Pascal buys Langhac mansion, near the abbey in Clermont.	Cyrano de Bergerac is born. Jean-Baptiste Colbert is born.	Jamestown, Virginia, creates the first representative assembly in the Americas. Slaves first brought to the colonies.

	PASCAL	FRANCE	EUROPE AND THE NEW WORLD
1620	Birth of Gilberte Pascal, who marries her cousin Florin Perier.	Jean Picard, French astronomer, is born.	Plymouth colonists set out from England. Witchcraft trials begin in Scotland. Francis Bacon publishes the *Novum organum*. Bonesetting becomes a science. Thirty Years' War begins in Prague.
1623	June 19: Blaise Pascal born in Clermont, the son of Étienne Pascal, a minor noble and government official, and Antoinette Pascal, née Begon.	Erotomania first mentioned as a mental illness. Phillipe de Mornay dies.	August 6: Urban VIII Barberini elected pope. Would order the trial of Galileo. Wilhelm Schickard invents the calculating clock, a first attempt at a computer.
1624		Cardinal Richelieu becomes first minister of France. Jean-Louis Guez de Balzac publishes his *Lettres*.	George Fox, founder of the Quakers, is born in England. War between England and Spain. Mail service begins in Denmark. Saint Rosalio makes a miraculous appearance at a plague in Sicily.
1624–1634		Richelieu builds the Palais-Royal in Versailles.	Cornelius Drebbel discovers gases.
1625	Birth of Jacqueline Pascal.	Henrietta Maria, princess of France and Navarre, marries Charles I of England. Thomas Corneille is born.	James I of England dies.

	PASCAL	FRANCE	EUROPE AND THE NEW WORLD
1626	Antoinette Pascal dies.		Saint Peter's Basilica is consecrated. Charles I dissolves Parliament. The Dutch settle Manhattan.
1627		Richelieu sets out to establish the supremacy of the crown.	The aurochs are hunted to extinction, with the last one killed in Poland.
1628		Richelieu defeats the rebellious Huguenots.	
1628		Richelieu founds the Académie Française.	
1630		Day of Dupes.	
1631	Étienne moves to Paris and directs his children's education based on the pedagogy of Montaigne. Blaise proves to be exceptional at mathematics.	The bell "Emmanuelle" in Notre Dame Cathedral is recast. René Le Bossu, French critic, born.	Imperial troops massacre about twenty thousand people in the city of Magdeburg.
1633		Saint-Cyran appointed father confessor to the nuns at Port-Royal de Paris by Mother Superior Angélique Arnauld.	Trial of Galileo in Rome. Samuel de Champlain, at the behest of Cardinal Richelieu, reclaims his role as commander of New France.
1635	Young Blaise revealed as a mathematical prodigy.	The Académie Française of Paris expands to become a national art society.	

	PASCAL	FRANCE	EUROPE AND THE NEW WORLD
1636		First performance of Pierre Corneille's play *Le Cid*. Jacques Marquette, French Jesuit and explorer, born.	Harvard College founded in the English colony of Massachusetts. The first ancestors of John Adams migrate to America. Roger Williams founds Rhode Island.
1637		René Descartes publishes the *Discourse on Method*.	
1638	Étienne goes into hiding after opposing a fiscal measure of Richelieu's, but leaves the children in Paris.	Louis XIV is born. Richelieu has Duvergier de Hauranne, the abbé de Saint-Cyran, imprisoned at Vincennes for the disruption of the peace of the church. The solitaries at Port-Royal move out to the old monastery at Port-Royal des Champs. The nuns remain in Paris. Cornelis Jansen dies. The French admiral d'Estrées runs his entire fleet aground in Curaçao.	Dutch settle in Ceylon. Maria Theresa of Spain, future wife of Louis XIV, born in Madrid.
1639	Blaise's sister Jacqueline appears in a play before Richelieu, after which he not only pardons Étienne but appoints him tax collector at Rouen.	Jean Racine, French dramatist and Jansenist, born.	Connecticut's first constitution, "The Fundamental Orders," is adopted. First printing house in the United States is founded in Cambridge, Massachusetts. Montreal settled.

	PASCAL	FRANCE	EUROPE AND THE NEW WORLD
1640	Pascal family moves to Rouen. Blaise publishes his short work *Essay on Conic Sections.* Essay later discussed by Leibniz.	Posthumous publication of Jansen's *Augustinus.*	First book, the Bay Psalm book, printed in America.
1641	Gilberte Pascal marries Florin Perier.		
1642	Birth of Étienne Perier, who will later confirm the genuineness of Pascal's *Memorial.* Blaise begins to work on his calculating machine, the Pascaline, to assist his father in computing taxes.	Cardinal Richelieu dies.	1642 to 1651: English Civil War between Cavaliers and Roundheads. The Puritans close all theaters in England. Galileo Galilei dies.
1643	Blaise continues work on the Pascaline.	Louis XIII dies. Anne of Austria becomes regent.	Antoine Arnauld publishes *De la fréquente communion.*
1644			Torricelli conducts his experiments on the vacuum.
1645	*Letter to the Chancellor,* dedicating the calculating machine.	Louis Joliet, French explorer of Canada, born.	English Civil War.

	PASCAL	FRANCE	EUROPE AND THE NEW WORLD
1646	Étienne Pascal and his friend Pierre Petit re-create Torricelli's experiment on the vacuum. Blaise takes over the experiments. Étienne is injured and is cared for by two Jansenists who convert the family to this strict form of Christianity. April 5: Birth of Margaret Perier. Blaise Pascal begins work on the vacuum.		The Westminster Confession of Faith. Wars of the Three Kingdoms.
1647	Pascal returns to Paris for his health. Jacqueline attends to him. Visits by Descartes on September 23 and 24. Discussion on atmospheric pressure and the function of the barometer. Controversy with Père Noël the Jesuit plenist and teacher of Descartes over the authority of Aristotle. Birth of Marie Perier.	Pierre Bayle, French philosopher, born. Denis Papin, French inventor, born.	Death of Torricelli.

	PASCAL	FRANCE	EUROPE AND THE NEW WORLD
1648	Conversations between Blaise Pascal and Monsieur de Rebours at Port-Royal. Much misunderstanding. Pascal returns to Clermont. Writes treatise on conic sections.	Sept. 1: Père Mersenne dies. Sept. 19: Florin Perier, following Blaise Pascal's detailed instructions, performs the great experiment on the Puy-de-Dôme. Blaise repeats experiments at the bottom and the top of the St. James tower in Paris, as well as in a tower at Notre Dame. Pascal considers the existence of the vacuum to be proved.	Peace of Westphalia; end of Thirty Years' War in Germany.
1648–1653		The Fronde of the Parlement. Revolt by the Parlement and nobles against the regency.	
1648		May: Chamber of St. Louis in Paris draws up demands for reform. August: Arrest of Broussel; Parisians rise up against Séguier and the queen.	
1649	Pascal family returns to Clermont to avoid the Fronde.	January: Revolt of the Parlement of Aix. The *frondeurs* and the French government sign the Peace of Rueil. Jansen's *Augustinus* denounced at the Sorbonne.	January 30: Charles I of England is beheaded. September 2: The Italian city of Castro is destroyed by the forces of Innocent X.

	PASCAL	FRANCE	EUROPE AND THE NEW WORLD
1650	.	René Descartes dies in Sweden.	Christopher Scheiner, great Jesuit antagonist of Galileo Galilei, dies.
1651–1653		Rise of the Ormée movement in Bordeaux.	
1651	September 24: Death of Étienne Pascal. September 27: Birth of Louis Perier. The duc de Roannez appointed governor and lieutenant general of Poitou.	Jean-Baptiste de La Salle born.	Thomas Hobbes writes *Leviathan*. Massachusetts passes laws forbidding poor people from adopting excessive styles of dress. Maximilian I, elector of Bavaria, dies.
1652	Jacqueline enters the convent at Port-Royal de Paris. Pascal begins his "worldly" period. Letter from Pascal to Queen Christina of Sweden.	Michel Rolle, French mathematician, is born.	Cape Town, South Africa, founded. Rhode Island passes the first law in the Americas against slavery. John Cotton, founder of Boston, dies.
1653	Pascal takes journey to Poitou, accompanied by Méré, Mitton, and the duc de Roannez. Pascal writes on the geometrical and intuitive minds.	Jews allowed to return to France and England. The Fronde ends.	February 2: New Amsterdam, later renamed New York City, is incorporated. Coffeehouses become popular across Europe.

	PASCAL	FRANCE	EUROPE AND THE NEW WORLD
1653	June 3: Five propositions of Jansen found in the *Augustinus* condemned by Pope Innocent X. June 5: Jacqueline Pascal takes vows at Port-Royal, taking the religious name of Sœur Jacqueline de Sainte-Euphémie.		
1654	November 23: a two-hour ecstatic vision leads to Blaise's conversion. The account of this vision is kept in the lining of his coat at all times.	June 3: Louis XIV crowned at Rheims.	The Republican Party questions Cromwell's power. Cromwell expels his enemies from Parliament. December 27: Jacob Bernoulli born in Switzerland.
1655	January 7: Pascal takes a retreat to Port-Royal, where he defends Arnauld against the Jesuits who sought to expel him.	Arnauld publishes his attack on Jesuit casuistry in his *Lettres à un duc et pair.* Motion to expel him from the Sorbonne.	Battle of the Severn. Protestant militia defeats Catholic militia for control of Maryland.
1656	Appearance of the first of the *Provincial Letters.*		The pendulum clock invented by Christian Huygens.
1658	Pascal lectures on his apologetics to the leaders of Port-Royal.		September 3: Oliver Cromwell dies.

	PASCAL	FRANCE	EUROPE AND THE NEW WORLD
1659	Pascal comes down with the illness that will lead to his death. Works in brief periods of relief from suffering.		
1661	Jacqueline dies. Port-Royal closed after official condemnation of Jansenism.	Death of Cardinal Mazarin. Priests are required to sign a formulary against Jansenism.	
1662	August 17: Blaise Pascal dies in the house of one of his sisters.		
1670	Publication of the *Pensées*, which Pascal had worked on sporadically the last four years of his life.		

The Man Who Played Dice with God

Each of us earns his death, his own death,
which belongs to no one else
and this game is life.

—George Seferis (Giorgos Stylianou Seferiades)

The real trouble with this world of ours is not that it is an
unreasonable world, nor even that it is a reasonable one. The
commonest kind of trouble is that it is nearly reasonable, but
not quite. Life is not an illogicality; yet it is a trap for logicians.
It looks just a little more mathematical and regular than it is;
its exactitude is obvious, but its inexactitude is hidden; its
wildness lies in wait.

—G. K. Chesterton

I am convinced that He [God] does not play dice.

—Albert Einstein

The most beautiful thing we can experience is the mysterious.
It is the source of all true art and science.

—Albert Einstein, *What I Believe*

This is the story of Blaise Pascal, the man who invented the modern world, or at least a good chunk of it. He lived thirty-nine years of the seventeenth century, and was perpetually sick. From childhood on he was in pain every day, but along the way he invented one of the first calculating machines, the very first public transportation system, probability theory, decision theory, and much of the mathematics of risk management, and proved the existence of the vacuum—all of which set the stage for quantum physics, the insurance industry, management science, racing forms, the computer, Powerball lotteries, Las Vegas, the vacuum pump, the concept of outer space, the jet engine, the internal combustion engine, the atomic bomb, mass media, and on and on. You cannot walk ten feet in the twenty-first century without running into something that Pascal's thirty-nine years of the seventeenth century did not affect in one way or another.

Pascal was also a religious mystic. His sister Jacqueline, who became a Jansenist nun at the Benedictine abbey of Port-Royal, played a major role in his spiritual life. She was also an early feminist, who argued vehemently for the right of self-determination for women. Later in his life—on Monday, the 23rd of November, 1654, the feast of St. Clement, pope and martyr—after enjoying a serious bout of worldliness, which Jacqueline chided him for, he sat alone in his room, buried in depression, and suddenly, from half past ten in the evening until half past twelve, he had a direct encounter with God that changed him. *Fire!* he said. *Certitude. Certitude. Joy. Peace.* He told no one about this, but he wrote it out as a short poem, a memorial of the event, and pinned it to the inside of his coat, next to his heart. No one would ever have known about it, except that his nephew found it after Pascal died.

Like his family and most of his friends, Pascal was a Jansenist, a member of a steel-rod sect within Catholicism that was declared a heresy in his own lifetime and that demanded a life of penance—if not for your own sins, then for the sin of Adam, which you inherited when your parents had too much fun conceiving you. Jansenism, like the more hard-core

Calvinism, followed Augustine in laying the sin of Adam onto the shoulders of every human born of a mother and a father, a sin that was passed on through the pleasure of sex, like an STD. And because of that sin, the human race has forever been wretched and vile, so that only the very few, those selected by God, can be saved. This was called God's mercy.

Amazingly, this variant of Christianity produced wave after wave of ferocious Christians, who in their zeal saw themselves as a holy remnant in an increasingly sinful church. Pascal suffered under this vision, received mystical insights under it, and wrote his greatest works, the *Provincial Letters* and the *Pensées,* in defense of it. However, in the middle of this dry, rigorous plain that he found himself crossing, he retained enough puckish sense of humor to invent a proof for the rationality of faith based on gambling.

This is a book about his life and how that life led to such a proof. Many people have tried to understand this man, to psychoanalyze him and thereby put him on the shelf as a garden-variety neurotic. Others have tried to canonize him, to see in him the marks of sanctity that he yearned to find in himself. He was neither a neurotic nor a saint. A faithful Catholic, he spent his adult life deeply involved with a heretical group that caused no end of trouble and that in fact so weakened the church through constant nattering argument that it was nearly powerless in the face of the French Revolution. As a scientist, he argued ferociously against the Aristotelian orthodoxy of his day, and many historians would hold that his arguments laid the foundation for some of the most important concepts of empirical science. So, how do we reconcile the scientist and the mystic? I don't think we can, and that is what makes Pascal so interesting.

His great discovery concerning probability emerged from one of those silly moments in life that often produce big ideas. On a trip to Poitou, while hanging out with his friend the duc de Roannez and his entourage, he stood in a small crowd of courtiers watching two ne'er-do-well gentlemen, the self-proclaimed legend the chevalier de Méré and his sidekick, Monsieur Mitton, gamble their hearts, not to mention their fortunes, away. Up until that point, the two sparkling gentlemen thought little of the short, brittle-looking fellow who seemed more interested in theology than in the games of real men, and they often laughed behind his back. They knew that their host, the duke, was quite fond of the little man, who always seemed to have a runny nose or a headache, and they knew vaguely that he had a reputation as a mathematician and a scientist, but what use was he beyond that?

Like two high school cool boys, they joked about him when he wasn't around and gave him nicknames that would have shamed either of them to receive in turn. Had Pascal been anyone else, he might have challenged one of them to a duel. Those would have been the rules for true gentlemen, of course. But Pascal was far too sickly to follow them. Then, after a long streak of bad luck, the chevalier noticed Pascal standing nearby and, remembering that he was a mathematician, asked him a question: if I am playing dice with Monsieur Mitton, and we agree to so many throws but our game is interrupted, how then do we divide up the pot?

They were in luck. Pascal was precisely the man to answer that question. Not only was he an accomplished mathematician, but he had also spent years looking for ways to apply mathematics to everyday life. He had actually invented one of the first computing machines, the Pascaline, when he was little more than a boy, simply to help his father, a tax judge, more easily complete the tidal wave of calculations that he was drowning under. Suddenly, Pascal wasn't quite the geek they had thought him to be. He promised them an answer, and soon was able to show just why the chevalier's strategy at dice was losing him so much money. A capital fellow!

What the gentlemen found after a time was a young man whose intelligence far outstripped their own, who was able to see further and think more clearly, who had a pungent wit, who could be cruel one minute and solicitous the next. They came to know a terribly unhappy young man who wanted to be a saint and yet loved life, who berated himself for his worldliness one minute and then laughed at some raucous joke the next. If they had been told that this Pascal fellow would soon have a mystical experience of God and withdraw from the world, they would not have been surprised, for he was halfway there anyway. If they had been told that he would soon after write one of the greatest and nastiest works of the modern age, the *Provincial Letters,* a book of satire that Voltaire kept by his bedside every night and used as a model, and that these letters would be directed against the all-powerful Jesuits, they would not have been surprised by that, either.

The Witch

Historians are the best gossips.
—John Padburg, S.J.

It seemed certain that the boy would die. Mysterious childhood ailments abounded, but this one was mysterious indeed.
The boy Blaise was only two years old, the first and only son of Étienne
and Antoinette Pascal, when suddenly he began to waste away, becoming
emaciated, as one *en chartre,* or "starving." He seemed dejected. He could
not stand the sight of liquids, nor could he take water in any form. In
fact, he seemed afraid of water, obsessed by a sudden hydrophobia that
set him shrieking with fear. What's more, he couldn't bear to see his parents together. His mother, by herself, was fine, as was his father, but the
two of them together sent him into rages. Was he possessed? Was he be-
witched? It was the seventeenth century, and most of the Pascals' fellow
townspeople would have thought such things possible. Rumors about the
boy and his illness fluttered about Clermont like birds.

The boy's father, Étienne, was uncertain. He was too much the scientific intellectual to easily believe in witchcraft and was, by his own account,
an *honnête homme,* a man of good breeding, one of the new bourgeois intellectuals who served his king and his God, and who made a little money

on the side doing it. He was a worldly man, though pious; a rational man; a philosophical man who doubted all the superstitious frippery of the simple people. But still, the boy was wasting away, and if the father did not find a cure, and find it soon, his son would die.

The town gossips suspected first this one, then that one, finally culling out an old woman who had once worked for the Pascals, possibly as a *sevreuse,* one who took in the children of the wealthy during their time of weaning, one who would put up with the children's tantrums and their weeping, one who, because of her age, could no longer be a wet nurse and who was therefore the child's first teacher about the hardness of the world.[1] This particular old woman had once received the Pascals' charity—a fact that, oddly enough, became the source of her grievance against Étienne, the tax judge. Because of her prior relationship, she had expected a favorable judgment by monsieur the judge, but was disappointed and grumbled about his hard-heartedness. And so the people put together the pieces. An old woman, a grievance, a mysterious illness—it had to be witchcraft. Étienne stopped her in the street one day and told her that if she was indeed responsible, he would take her to court and see her punished. He demanded that she cure his son at once and without further witchcraft.

The old woman, cowed, apologized over and over, and said that certainly she would do what she could. Still, life for life, death for death: some other poor soul would have to sacrifice its life for the boy. After all, the spell had been a killing spell, which could be satisfied only by a death. Pascal ridiculed her and asked if she wanted one of his horses to kill, but the old woman pushed on. She said no; a cat would do as well. In time, the bargain was set. Étienne returned to his home, and the witch cast about for a cat to steal. She went out and found herself a cat. But as she was walking up the stairs of Étienne's house, she met someone on the stairs, likely a servant, who opposed her, insulted her, and upbraided her. Startled and upset, she threw the cat out the window. Now, cats are legendary when it comes to surviving falls, and the window was not very high off the ground. But when the old woman found the cat outside the window, it was already dead.

To complete the cure, the witch then gathered common herbs from the garden and, after mixing them with flour, placed them on the boy's navel. Suddenly, little Blaise fell into a coma and looked as if he had fallen dead, just like the cat. Étienne called for the doctor, who arrived and examined the boy, and then told the distraught parents that their son had indeed died, that he was sorry, that there was nothing he could do. Meanwhile, the witch had gone off for a time, but after a while she returned. She knocked on the door, and the servants ushered her into the child's bedroom. Overcome with grief and anger, Étienne the philosopher, the gentleman of good breeding, ran to the woman and knocked her to the ground with his fist. Standing over her, he shouted at her and cursed her. But the witch pleaded, assuring him that his son was not dead, that they should not put him in a shroud, and certainly not bury him—that this lethargy was part of the cure, and that if they only had a bit more patience, they would see that Blaise would awaken soon and begin to heal.

The spell required that they wait until midnight, when suddenly, she said, the boy would awaken and return to himself. That afternoon and into the night, the Pascals—Étienne and Antoinette—with their few servants and possibly even their daughter Gilberte, the woman who would eventually write down the story as part of a biography of her brother, Blaise, stayed by his bedside and prayed, taking cold comfort in what the witch had told them—that the boy was not dead but only asleep. Midnight passed, and nothing happened. One o'clock, and still nothing. Two o'clock. Three, four, five, six. The family despaired, but then around six thirty in the morning the boy stirred, and finally his eyes fluttered awake. The first thing he saw was his father and mother next to the bed, standing together, and he began to cry, as he had done in the past. So they knew that he was not yet cured. Moreover, he still feared water. But after a week, Étienne returned home one evening to find Blaise sitting in his mother's lap, pouring water from one glass to another. He tried to approach his son, but Blaise began to cry. This situation continued for another few days, when Étienne found them once again, mother and child, and approached, but this time Blaise did not object, and began to put on weight from that point on, until he looked as if he had never been sick.

When pressed, the witch admitted that she had placed the spell on the boy after Étienne had refused her application, and when pressed further, she admitted that her suit had not been just and that she had hoped her previous association with monsieur the judge would get her what she wanted anyway. The ways of the Evil One are indeed slippery, the people told themselves, and were satisfied. What happened to the old woman after that has not been written down.

But what was this terrible illness? What had nearly killed the boy? The symptoms when taken together make little sense and have to be treated separately. The fear of water and the alarm at seeing his parents together might be psychological in nature, an Oedipus complex in the making. The wasting, together with the deformation in the child's bones and skull, are easier to diagnose as a form of marasmus, a childhood disease that arises from a deficiency of protein in the diet and often occurs today in developing countries soon after a child is weaned. If the local diet is deficient in protein, if it relies mainly on cereal grains, with a lack of trustworthy cow's milk, marasmus is common enough. This was the case in seventeenth-century France, especially in the Auvergne region before pasteurization, where cow's milk was routinely turned into cheese, thus concentrating the fat and resulting in a loss of much of its protein content. All of Europe was in the latter days of the Little Ice Age, which stretched from the fourteenth century to the eighteenth century. During this time the Gulf Stream stopped shipping warm water from the Caribbean to the coasts of England and France, and the climate of Europe chilled enough to induce regular crop shortfalls and even sporadic famines. The European diet in the seventeenth century was meager at best—bread and wine mostly, and cheese, with a few vegetables. There was wheat gruel for the children, but only a rare piece of meat, even in the better families. A child with a delicate condition might well have been unable to absorb what few proteins were to be found in his food.[2]

It is likely that Blaise's condition began soon after his mother ceased to breast-feed him. Perhaps she wanted another child, and since many women used breast-feeding as a means of birth control, Antoinette Pascal may have decided that Blaise was ready to make that fearful transition from a diet of breast milk, which had everything he needed to survive, to the local diet, which often left adults hungry. Weaning was sometimes a life-threatening experience for a child, the second great shock in life after birth itself. Gastrointestinal infections followed marasmus like jackals following a lion, and children often died of dysentery, wasting away. The food did not satisfy, and the only thing a child of two knew was that he used to be fed and now was hungry, and that the presence or absence of his mother made the difference. Perhaps this explains some of Blaise's anxiety at seeing his parents together—perhaps he recognized in his toddler's brain that a sea change had occurred in his world, a change that pushed him beyond anything he had known into a strange new situation.

Life for Blaise Pascal began with uncertainty. His sister Gilberte reported how when the doctors cut open his body after his death, the autopsy showed that the anterior fontanel, the "soft spot" that opens in early childhood and then closes as the skull develops, had opened once again in Blaise's case, and that it had not closed but had filled in with softer cartilage. The doctors who performed the autopsy said that this was because of Pascal's native genius—because his brain, which was too large for his skull, had forced the fontanel to open and would not let it close. The fact that Pascal suffered debilitating headaches through most of his life and that his health had been precarious from childhood on, they assumed, supported their explanation, and was the price one paid for a prodigious intellect. However, these were also signs of childhood starvation and were symptomatic of rickets, a disease caused by a deficiency of vitamin D and of calcium. Perhaps genius begins with deprivation.

A Dangerous World

Arise, ye prisoners of starvation,
Arise, ye wretched of the earth,
For justice thunders condemnation—
A better world's in birth.

—"The Internationale" (1871)

The boy Blaise lived, but he was sickly most of his life. He suffered debilitating headaches, spells of exhaustion, leg paralysis, stomachache, toothache, and beneath it all a grinding melancholy. The fact that a child of such a prominent family as the Pascals should suffer from malnutrition says something about the realities of the time. His family, though not wealthy, was more than respectable, having roots in the minor nobility. Étienne had been born out of two strands of the Pascal family, for both his mother and his father were Pascals. Blaise's grandfather had been a commoner, for that branch of the family was never quite able to hold high office long enough to be able to pass nobility on to their children. Blaise's grandmother's branch, however, did have a few teetering instances of gentry, and so Étienne, and therefore Blaise, could claim some aristocrats in his bloodline.

In 1610, Étienne Pascal bought his way into an office, a common enough practice, and served as a lawyer on the *généralité,* one of the less prominent tax boards. Cardinal Richelieu had recently become the first minister to King Louis XIII, and he would soon raise the crown over all other powers in France. As public servants, the Pascals would be part of this conquest and would benefit from it. In 1624, the year after Blaise's birth, Étienne became second president of the Cour des Aides, a court dedicated to collecting often burdensome taxes for a series of spendthrift kings, and to settling any disputes that might follow. If the Pascals could have held this position for three generations, they would have been permanently ennobled and become part of the *noblesse de robe,* the nobility by virtue of governmental service—as opposed to the *noblesse d'épée,* the old aristocracy, the nobility by virtue of the sword. Étienne was therefore a representative of the crown, a bureaucrat administering royal power, and although this was not too great a problem in the Auvergne, the region of Clermont-Ferrand, it would later become a dangerous business when Cardinal Richelieu transferred Étienne to Rouen, which was in revolt over the weight of the king's taxes.

By all accounts, Étienne Pascal was a man of the world more than he was a man of the faith, a man of science and of philosophy more than religion. He was an amateur mathematician, one of the most eminent in France, who loved the life of the mind. He came from that class of men whom the Jansenists, including his own son, Blaise, would later accuse of being laxists, of being soft on sin and soft on sinners. This was the same crowd that would eventually produce the *libertins érudits,* the scoffers and doubters, who would later become Deists and agnostics—that class of men who would eventually lead the Enlightenment and, finally, the French Revolution. In Étienne's day, however, most of the intellectuals of France were not willing to criticize the church, except as humanists might—as loyal participants in Christianity who criticized the church only as a means of helping her find her true self. They were therefore the heirs of the tradition founded by Erasmus and Thomas More a century earlier.

In 1616, Étienne Pascal married Antoinette Begon, who was twenty, the daughter of a Clermont merchant. They had four children: Gilberte, born in 1620; Blaise, born in 1623; Jacqueline, born in 1625; and another child, Anthonia, who did not survive. In 1626, when Blaise was three years old and Jacqueline only a toddler, Antoinette died. No one knows how or why she died—probably from one of the many bouts of plague that brawled through the city. Little is known about her; little has been said. The only testimony we have of Antoinette's life is from Gilberte's daughter Marguerite, who noted that her grandmother was a pious young woman devoted to charitable works. Sadly, Antoinette died while her children were still young enough to have only an impression of her, and not much memory.

We can only guess how the mother's death altered the relationships between the bereaved father and his three remaining children, but we do know that Étienne was transformed by the loss from an ordinary careerist in the French bureaucracy into a man utterly dedicated to the education of his children. The children relied more and more on a father who grew ever more protective, especially of his son, Blaise, whom he would not allow to grow to become an independent adult.

Perhaps Étienne's resolve to retire from his career and dedicate himself to his children was the only possible answer. Perhaps, for an honorable bureaucrat, it was the greater sacrifice. To place them in the care of a nanny and then distance himself from their lives for the sake of his work—that would have been the normal thing to do. But at this point, Étienne Pascal did something unusual for the time: he refused to remarry. Perhaps his grief at Antoinette's death was still too great. Whatever the reason, a year later Étienne gathered his children and moved them all to Paris. Whether he did so to flee his own grief or to pursue a new life for himself and his children, he left all that he had ever known behind. He separated his children from the day-to-day life of their cousins and aunts and uncles, isolating them even more from family and friends, binding them even closer to him. Such a move would have been from grief to isolation, from a provincial city where everyone knew everyone else's business to the great city where no one really knew anyone. It is not surprising, then, that toward the end of his life, Blaise would write in his *Pensées*: "We want truth and find only uncertainty in ourselves. We search for happiness and find only wretchedness and death."[3]

All things considered, Blaise Pascal was born into a dangerous and changing world, be that the world of the Auvergne or of Paris. His birthday was June 19, 1623, the feast of St. Romuald, a Benedictine monk and founder of the Camaldolese order of hermits, a man who fought the devil all his life. There were plenty of devils afoot, even in Pascal's day, Protestant and Catholic alike, and uncertainty was king. Copernicus had turned the universe inside out, and scientists were trumpeting this all over Europe. France had not yet recovered from the Huguenot wars—where Catholic and Protestant took arms against each other. Though victorious, Catholic France was wounded, hobbling through the early seventeenth century on the stump of victory. The Edict of Nantes had ended that phase of the religious wars, but it was only twenty-five years old, and people were still alive who remembered the massacres and the streets that flowed with blood. The Catholic world that Blaise Pascal knew was anxious, uneasy, troubled.

After the earthquakes of Luther and Calvin, French society, like much of the rest of Europe, had split in two. *Une foi, un loi, un roi*—one faith, one law, one king. This was the cry. And yet, with the coming of the Reformers, the medieval unity quickly broke down, and the nation broke with it. There were seven Huguenot wars in total, short but vicious, running from 1562 until 1580, years marked by assassinations and massacres. The worst of these occurred on St. Bartholomew's Day in 1572. Some of the old people in Étienne Pascal's circle may have been there; some may have participated.

The Catholics and the Huguenots had already fought two wars, followed by the Peace of St. Germain, when the aristocracy kissed and made up, and offered their sons and daughters to one another in marriage to cement the deal. Catherine de Médicis had worked tirelessly to bring the two sides together. Jeanne d'Albret, the queen of Navarre, offered her daughter Marguerite to the Huguenot Henri de Navarre, and even Elizabeth I of England contemplated marrying a younger brother of the Catholic King Charles of France. They were all ready to make peace, but no one included the common folk in their plans.

While the aristocrats sent envoys to each other, the tension on the streets multiplied. Huguenot sermons started calling for civil disobedience against the Catholic king. Even Jean (or John) Calvin, who traditionally advised his followers to obey the king, changed his mind and preached that because the Catholic King Charles and his court were disobeying God by remaining Catholic, no good Calvinist owed them his allegiance.

Rumors of a mythical constitution of the ancient Franks, revealing how the old kings were chosen by a vote of the people and ruled by their advice and consent, appeared in a popular tract, *Francogallia,* by François Hotman. Thus, in the minds of the Catholic French commoners, the Protestants had begun preaching republicanism—treason to the crown— to add to their treason against God. The Huguenots were now destroyers, spoilers, revolutionaries against the harmony that had once been France. The fact that this unity the Catholics looked to was as mythical as the old Frankish constitution never came up. Then France fell into a recession, and once-prosperous people found themselves hungry.

On a hot summer night, when riot was buzzing in the taverns and the hotels, someone took a shot at Admiral de Coligny, the leader of the Huguenot military. He was wounded and nearly killed, but not quite. The Huguenots demanded justice and appealed to the king. When justice did not arrive at once, they threatened to riot. But the king, Charles IX, and his court, which included Catherine de Médicis and Henri, the duc d'Anjou, struck first. The king had been uncertain initially, fearing, quite justly, the general slaughter that might follow, but the others pressured him until he agreed. They met in secret on August 23 and decided to assassinate Coligny. Charles waffled even then, but after some discussion he finally surrendered, saying, "Well, then, kill them all, so that no one will be around to reproach me after."

Early that morning, a Sunday morning, the duc de Guise led several troops of Swiss mercenaries and a few French regulars to the admiral's home. It was the king's will and God's will that they should take their vengeance on the traitors and rebels who had fallen into their power, he told his men. Since the soldiers were not in uniform, they identified themselves with a white armband on their left arms and a white cross stuck in

HOLLY TREE FAMILY PRACTICE, P.A.
LARRY A. BERGLUND, M.D.
DEA # BB 1718393 LIC. # 14433(SC)
LINDA L. GIAMBALVO, M.D.
DEA # BG 3240506 LIC. # 16718(SC)
MARK T. WHITE, M.D.
DEA # BW 1975931 LIC. # 15006(SC)
WILLIAM J. TAYLOR, M.D.
DEA # AT 2811708 LIC. # 12368 (SC)
KELLIE DAVIS-HALL, F.N.P.
DEA # MD 1691737 LIC. # 3230
FAMILY PRACTICE
1338 HIGHWAY 14
SIMPSONVILLE, SC 29681
(864) 297-7091

NAME _____ AGE _____

ADDRESS _____ DATE _____

℞
TAMPER-RESISTANT FEATURES INCLUDE:
SAFETY BLUE ERASE-RESISTANT BACKGROUND
AND "ILLEGAL" PANTOGRAPH

Othello's
4 pain

Refill _____ times

Label

DISPENSE AS WRITTEN _____ SUBSTITUTION PERMITTED _____

8HFP503345B9

their hats. They had all agreed that once the admiral was killed, the signal for the massacre would be given: a quick toll of the palace bell.

At gaining entry into the admiral's house, the soldiers rushed in, slaughtering the servants as they passed. They came upon the admiral in his room. Still believing he had the king's favor, he had refused to leave the city and had been awakened in the belief that what he was hearing in the street was another quick uprising among the people, something that the king's soldiers would quickly put down. Fully awake, he had left his bed and was praying when the soldiers rushed into his room, dragged him from his prayers, and stabbed him over and over. The duc de Guise, who was waiting below in the courtyard, called to the men in the house to find out if the deed had been done. "It is done," one of the captains called down, and threw the body out the window. Several of the men, including the chevalier d'Angoulême, backed away in disgust at the butchery, but the duke laughed at them, repeatedly kicked Coligny's dead body in the face and in the groin, and said to the chevalier that he and the men should cheer up. Since the king had commanded it, they should do the deed thoroughly.

The duke then ordered that they give the signal, that their fellow in the palace should ring the bell. All around them, voices cried out in the night, "Alarm! To arms!" The soldiers then dragged the admiral's body to a nearby stable and tossed it inside, where they decapitated and then eviscerated it, spitting on the parts and kicking the head around the stable like a football. Later, a mob of children came upon the body and tried to throw it into the Seine, but others stopped them and instead hung the headless admiral from the gibbet of Montfaucon.

Faster than light, news of the murder spread throughout the city. The local militia rose up, followed by the populace, who, in imitation of the soldiers, stuck white crosses in their hats to identify themselves as Catholic, and turned on the Protestants. Neighbor slaughtered neighbor, and all the bile of religious hatred poured out onto the streets to mix with blood. The killing continued for three days. It spread like plague from the city out to the surrounding towns and villages, out to the provinces, to the wineries, to the little dairy farms, to the fields of lavender. Many people believed that they had been ordered by the king to kill all the Protestants, and they set out to do just that. By the end of the third day, over seventy thousand people had been slaughtered.

Though the Huguenot wars had ended with the Edict of Nantes, the religious wars carried on. Twenty-five years later, in 1623, the year Blaise entered hungry into the world, the second phase of the Thirty Years' War ended. The Dutch once again fought the Spanish for their independence, and lost. That same year Pope Urban VIII, the pope who would command the inquisition of Galileo, would be elected to the throne of St. Peter; the Safavid Turks would conquer Baghdad; and "erotomania"—the delusion that a person, usually of a higher social station, is secretly in love with the delusional person—would officially be defined as a mental illness in Jacques Ferrand's *Maladie d'amour ou mélancolie érotique.* Wilhelm Schickard would build the first calculating clock that year, and the Plymouth Colony would celebrate its second Thanksgiving.

Two years later, in 1625, the year of Blaise's childhood malnutrition, the third phase, the Danish phase, of the Thirty Years' War would begin. Nations opposed to the Hapsburgs—the French, under the command of Cardinal Richelieu, because he feared Hapsburg power and the power of Spain; the English; and the Dutch—formed a league and handed over control of their armies to the Danish king, Christian IV, who held vast sections of northern Germany. The Danes invaded, but the Catholic League, under the command of the ambitious, and ruthless, General Albrecht Wallenstein, the patron of Johannes Kepler, crushed them irrevocably.[4]

In such a world, average French citizens lived as if by the roll of the dice every day. Would some king or his general fall on them that day and slaughter them all in the name of God? Would the crop be adequate in that harvest, the weather indulgent? Would God let them and their children live one more day? How could anyone understand, let alone live with, such uncertainty?

A Thinking Reed

Be not afraid of life. Believe that life is worth living,
and your belief will help create the fact.

—WILLIAM JAMES

To conquer without risk is to triumph without glory.

—PIERRE CORNEILLE

M an is but a reed," wrote Blaise Pascal in his last work, the *Pensées*, "the most feeble thing in nature, but he is a thinking reed. The entire universe need not arm itself to crush him. A vapor, a drop of water suffices to kill him. But, if the universe were to crush him, man would still be nobler than that which killed him, because he knows that he dies and the advantage which the universe has over him, the universe knows nothing of this. All our dignity then, consists in thought. By it we must elevate ourselves, and not by space and time which we cannot fill. Let us endeavor then, to think well; this is the principle of morality."⁵

Certainly it was a risk, an act of belief, perhaps even faith, for Étienne Pascal to cart his children all the way to Paris, so far from their family and friends. Likely, his main reason for doing so was to introduce them to the more cultured life of Paris, to the circle of great intellectuals, and possibly to a new life at court, where the intellectual world of a provincial

tax judge could expand beyond expectations. There is some evidence that
Étienne, seeing some intellectual promise in his son, Blaise, had developed a master plan to turn the boy into one of the great minds of the day.
He succeeded in this, even though Blaise himself later rejected that life
and embraced the rigors of Jansenism. Blaise was by Gilberte's account
a precocious little boy who asked questions far beyond his years and held
conversations that would seem appropriate to an adult. This, of course,
may be more mythology than fact, for it was common in seventeenth-
century France to describe saints in their youth in biblical terms, like Jesus
sitting among the doctors of the law in the Temple of Jerusalem, astound-
ing them by the acuity of his questions. Thus, sanctity and genius were
intertwined. In Gilberte's way of thinking, if Blaise had already become
a great man by the time of the writing of her biography of him, then,
like the saints who as toddlers wanted to listen only to the wisdom of
old men, he would certainly have shown that same kind of supernatural
promise, that divine gift, early in his life.

Étienne set about the task of educating his children, and along the way
created an innovative regimen of homeschooling. He proved to be quite
a capable teacher, a man who was ahead of his time in understanding the
psychology of education. It was his maxim, according to Gilberte, that he
would always keep his lessons at a level just above the level of the work
his students were capable of. Thus his children had to strive to understand
that which was in sight but which was just beyond their grasp. In this way,
Étienne built up the confidence of his children by giving them problems
that they could solve, but only with sweat. Each solution then became
another triumph and allowed their minds to grow, leaving them secure
in the knowledge that they could solve whatever puzzles were laid before
them.

Homeschooling, however, had its drawbacks, for Blaise was never allowed to attend school and thus never learned the art of fine negotiation
that most children learn on the playground. Thus he remained ignorant,
at least experientially, of many of the subtleties of human life. He never
attended school or matriculated at a university. He never married or
even seriously courted. Although Étienne introduced his children to the

intellectual life in a profound way, he failed to give them the kind of emotional training one needs to live a fully human life. Had Antoinette been alive, that might have been different.

One source of Étienne's pedagogical method was his own experience of mathematics—how it could become an all-consuming fascination that could distract the mathematician from other kinds of study. He was also concerned that, given Blaise's fragility, he not tax his son's strength. He therefore refused to allow Blaise to study mathematics until he was sixteen. He did not want him to be caught by this great passion too early, until he had been firmly grounded in grammar and in languages, especially the classics and classical literature. Instead, he presented his children with little problems in natural science. At the dinner table one evening someone struck a porcelain plate with a fork, and Blaise asked why the plate hummed. What was the cause of the sound? Why did the sound stop when you put your hand to the plate? After dinner, Blaise went about the house striking dishes with various kinds of silverware and found that different plates made different sounds, each with its individual pitch and timbre. In this one moment, Étienne introduced his son to experimental science, and encouraged him at each step. The problem, however, was that Blaise was in fact as precocious a child as Gilberte indicated. He was curious and, when given a boundary by his father, could not help but try to jump over it.

When Blaise was about eight years old, he spent much of his free time lying in front of the fire in his room, drawing diagrams in charcoal and working out calculations on the stones in front of the fireplace. He knew that he was breaking his father's rules against studying mathematics, and he tried to keep his work secret. At first, he tried to draw a perfect triangle, and then a perfect circle. As he came closer and closer to this, he began to develop his own language for his new geometry. He called a line a "bar," and a circle a "round," and, using his new vocabulary, he set about re-creating Euclid's ideas. He actually managed to reconstruct several of Euclid's theorems before his father walked in on him and found him drawing on the stones. Unseen, Étienne watched from a distance for a long time and then approached. Gilberte does not say who was more disconcerted at the discovery. Blaise had been caught disobeying his father's orders, but for Étienne it was a happy capture, for he found his son busy working on a project much beyond his level of maturity. Suddenly he realized that Blaise was not just precocious, but a prodigy. What could he do with such a son? What should he do? Both thrilled and fearful, nearly in tears, according to Gilberte he left his son alone by the fire, to continue re-creating the work of a man he had never read. Étienne said nothing about disobedience.

Blaise Among the Geometers

For he by geometric scale,
Could take the size of pots of ale.

—SAMUEL BUTLER

Philosophy is written in this grand book—I mean the universe—
which stands continually open to our gaze, but it cannot be
understood unless one first learns to comprehend the language
and interpret the characters of mathematics, and its characters
are triangles, circles, and other geometrical figures, without
which it is humanly impossible to understand a single word of it;
without these, one is wandering about in a dark labyrinth.

—GALILEO GALILEI

Fighting back tears, Étienne left Blaise to his studies and hurried to the house of his friend, another mathematician, a man named Jacques Le Pailleur. Once there, he wept openly, and Le Pailleur, concerned that some tragedy had fallen on the Pascals, fretted. What could cause his old friend to be so upset? Étienne stopped him mid-fret and told him that he was not weeping from grief, but from joy, and showed him some of the papers onto which Blaise had transferred his fireplace diagrams and calculations. After glancing at the handful of drawings, with symbols and arrows drawn in a child's hand, Le Pailleur realized that Blaise was a gifted child. He saw that, having been denied mathematics by his father's pedagogy, Blaise had simply invented it for himself.[6] Le Pailleur advised Étienne to abandon his course of study and to introduce the boy to mathematics at once. When Étienne returned home, instead of punishing Blaise, he presented him with a copy of Euclid, and told him to study it in earnest.

What better gift could a young intellectual have had at that time than a gift for mathematics? Mathematics was, after all, the royal science. The medieval universe was fading away, and the old divine certainties were losing ground. The scientists and philosophers of France were busy casting about for something new to bet their souls on, a new ground of order, a new way to make the universe spin properly, and for most of them, that something was mathematics. Everyone in France used it; it was the latest, hottest thing. Those in the inner circles of thought passed around treatises on geometry like junk novels at the beach, while merchants sought new ways to turn their business dealings into numbers.

Even the philosophers and theologians turned to mathematics for insight. The great French gardens were finger exercises in geometry; the vast, ostentatious *hôtels* of the high aristocracy were designed and built according to mathematical principles. Metaphysics, before and after Descartes, was gradually becoming a creature of mathematical logic. The pinnacle of reality was the pinnacle of order, and mathematics was the measure of that order. In their deepest hearts, French intellectuals thought that God was the ultimate mathematician, and now Étienne Pascal's own son had proved himself to be an adept at reading God's mind, a mathematical prodigy, a child who was born to geometry just as Mozart, 150 years in the future, would be born to music.

Sometime after, Étienne brought young Blaise along when he attended the little gathering of mathematicians and scientists that met in Père Mersenne's monastic quarters. Marin Mersenne was one of the great scientific majordomos of the age, a defender and promulgator of Galileo's astronomy, a gatherer of mathematicians and natural philosophers, and a great opponent of those mystic fakeries alchemy and astrology. He was a priest, a member of the Order of Minims, the most humble of all religious orders. Mersenne's little group, which would eventually become the Academy of Paris, the French equivalent of the Royal Society, would likely have met in his monastic cell sitting on hard, straight-backed chairs without much padding, in a circle or around a table. Some may have smoked tobacco in long clay pipes, since smoking was not forbidden—because, after all, better to tax it than forbid it, and because even Catherine de Médicis, the great queen of France from the century before, often took snuff as a cure for migraines. They would drink wine, not coffee (for coffee was a Calvinist drink, a Huguenot drink for the rising bourgeoisie, promoted by the Protestants as a way to awaken humanity from its Catholic alcoholic stupor to a new world of activity and industry). A bit of bread, a bit of onion, a bit of cheese, a bit of wine, and along the way they discussed matters of scientific merit, from methods of identifying prime numbers to new ways of marrying algebra with geometry to the failures and weaknesses of alchemy. Salacious, even radical, conversation they left to the libertines, those scoffers and doubters, those Deists, who were not welcome at Père Mersenne's table. They would leave such libertine conversation to the insidious salon of Madame Sainctot, the retired courtesan with the notorious past who was another friend of Étienne Pascal's.

This little group gathering in Père Mersenne's monastic cell made quite a splash. Some of the best minds in Europe were there. Descartes was a member. Mersenne himself had been the leading investigator into prime numbers. His formula $n = 2^p - 1$ (where p is a prime number) was not perfect for identifying primes, but it came close, and it is in fact still being used to help identify large primes. As a scholar, he was so connected, through letter and personal contact, with the leading thinkers of the time that many said that telling Père Mersenne about a new idea was the same as publishing it.

Pierre de Fermat, Blaise's future correspondent on probability, was also a member, and it was in Mersenne's monastic cell that Blaise first met him. Fermat is famous even today for his last theorem, and for his work on spirals and falling bodies.[7] He first came to Mersenne's group by writing to the priest and by correcting some of Galileo's titanic mistakes in geometry. He also developed new ways of determining the maxima and minima in an equation's curve, methods that conflicted with Descartes' own ideas, already published in his *La géométrie,* where he set forth his view of algebraic geometry. Needless to say, this set off a feud. Descartes wrote, expressing his dislike for Fermat's method for determining maxima, minima, and tangents, and Fermat fired back. Étienne Pascal entered the war on the side of Fermat, as did Gilles Roberval, a royal professor of mathematics and another of Mersenne's group. Descartes asked Girard Desargues, yet another member, to referee, and soon after he was proved wrong. Descartes had the good grace to admit it in a letter, though grudgingly. Nevertheless, those who had sided with Fermat were from that point on in a bad odor with René Descartes.

It was about this time that Desargues, a man known to both Pascals, published a book on conic sections that would be a strong influence on the young Blaise, leading to his first published work. Desargues' book had the unlovely title of *Brouillon projet d'une atteinte aux événements des rencontres du cône avec un plan,* or Rough Draft for an Essay on the Results of Taking Plane Sections of a Cone. Inside it, however, was an entirely new type of geometry, a projective geometry, with a new way of looking at conic sections as having properties that are invariant under projection. That is, if you draw lines through points on conic sections, those lines form projections out into the space around the conic section, and those projections will act in regular ways. In this way, Desargues invented a unified theory of conic sections, something that had not been done in that way before. This was quite an achievement, one that entranced the young Pascal, who had come to believe that geometry was the path to understanding the greatest truths of all. A short time later, after his family had moved to Rouen, and while the city was on fire with revolution, Pascal would make his own contribution to this new geometry. What better way to retreat from the violence of the outside world than into your own mind?

Un Bâtard Magnifique

Deceit is the knowledge of kings.

Qu'on me donne six lignes de la main du plus honnête homme,
j'y trouverai de quoi le faire pendre.
[Give me six lines written by the most honest man, and I would
find some reason there to have him hanged.]

Pour tromper un rival l'artifice est permis; on peut tout employer
contres ses ennemis.
[Deception is permissible to mislead a rival; use every means
against an enemy.]

—ARMAND JEAN DU PLESSIS, CARDINAL RICHELIEU

One must never forget that Étienne Pascal had taken his children to the Paris of Cardinal Richelieu—a fashionable place, a wealthy place, a place of power, where the Bastille glowered over all things. The Pascals had met the cardinal on several occasions, for they were themselves a fashionable family; they changed residences five times during their stay in Paris, always from one smart district to another. Their first house was on the rue de la Tissanderie, in a district where Henri IV had built two new palaces with their elaborate lawns and avenues. After a few years, Étienne brought his family across the river Seine to the Alberg-Charmaine, across from the great *hôtel* of the king's cousin, the prince de Condé, the man whose son would eventually lead a revolt against the regency of Louis XIV during the Fronde of the Princes, a coup d'état by the old nobility, misnamed really, for the term *fronde* refers to a popular uprising.

The cardinal was at the height of his power in those years. An odd man, diminutive, oppressed by his own conscience, sickly, he was a conservative churchman who nevertheless considered his first duty to be to his king and not to his church. He had set himself the task of welding that collection of feudal principalities that was France into one of the great nations of Europe, under an all-powerful absolute monarch, a monarch whom the cardinal controlled. Order was his abiding spiritual principle, and though he often used the rack as a finger of government, he used it no more than did any other ruler at the time.

Richelieu worked incessantly; he was never far from his desk, a fact that showed his middle-class roots. Armand Jean du Plessis de Richelieu, the cardinal, was a breathing contradiction. He had an iron will but a weak constitution, frail and sickly, pale under his ecclesiastical robes. Though perpetually ill or suffering from the fear of illness, he terrorized the entire court.

In spite of his frailty, once he was dressed in his cardinal's robes, his stern, unbending appearance forced people into submission. And he even exercised this power though Louis XIII, his great protector and benefactor, did not like him very much. Everyone knew this. As he was with so many people, Louis was outwardly courteous but cool. But every time he tried to oppose the cardinal, Richelieu appeared before him and presented his case, step-by-step. Richelieu was so rigorous, so rational in his argument, that Louis could not help but agree.

Various cabals gathered to rid the court of the little cardinal but failed. The man behind most of them was the king's brother Gaston d'Orléans, a wastrel addicted to gambling, good times, and irresponsibility, though he was his mother's favorite son. Time after time, in a bid for the throne he tried to unseat the cardinal in order to make his brother the king, who was chronically ill from tuberculosis, more vulnerable, for everyone knew that it was the cardinal who kept the king secure on his throne.

But Gaston was not the cardinal's only enemy. Both queens—Anne of Austria, the wife of King Louis XIII; and the Queen Mother, Marie de Médicis—detested him and wanted to be rid of him, and on at least one occasion they tried to unseat him. But he proved stronger than they, and in the end the Queen Mother, along with Gaston d'Orléans, had to flee France for their lives.

Richelieu had some experience, and some talent, in putting down revolts. He had a world-class intelligence service, led by a Franciscan monk, Father Joseph. In 1626, the monk's agents caught Gaston in a conspiracy. The cardinal could not imprison a member of the royal family, so he went after the other conspirators: the governor of Gaston's territory, Marshall Dornano; and Gaston's friend Henri de Talleyrand, the marquis de Chalais. Richelieu had them arrested, had the friend executed, and sent the governor to the Bastille, where he died. After that, one after another, the cardinal removed the teeth from his once-powerful opponents. The prince de Condé, the king's cousin and the Pascals' neighbor, a famous warrior and a lion on the battlefield, eventually held the curtains open for the cardinal to let him pass. Others did the same. Even the king was easily bullied, for Richelieu had built for himself the perfect place for any bureaucrat: he was indispensable. Louis was a weak, vacillating king, a deeply closeted homosexual susceptible to older men. Without much of a mind of his own, he left the job of running the country to his minister, Richelieu. The problem was that Louis had a conscience, an unfortunate possession for a king, and he vacillated between the realpolitik of Richelieu and the dictates of his own conscience. But Richelieu always seemed to win, even over the king's scruples, for he set such moral issues against Louis's own self-interests. And the fact that the cardinal's oppressive tax policies kept the money flowing didn't hurt, either.

Richelieu was an amateur at finance, but his policies, though not very successful in growing the economy, managed to wring enough money out of the people to keep the royal family in palaces. Everyone at court knew this and gnashed their teeth in frustration. Meanwhile, powerful men ended their lives in the Bastille or swinging at the end of a rope. "Give me six lines written by the most honest man, and I would find some reason there to have him hanged," Richelieu once said. Everyone knew this to be true.

Historians have been hard on Cardinal Richelieu, and perhaps unfairly. The Enlightenment pundits loathed him; their novelists and playwrights made fun of him, because he was the architect of the Old Order. Nevertheless, it would have been a disaster for France had the cardinal fallen. Gaston was too self-involved to be a good king, and he owed too many favors to the old feudal princes to be his own man. The same can be said of the prince de Condé, for he was the man in charge of the old baronial party. Had either of these men dethroned Louis and his first minister, France would have shattered into a hundred feudal pieces and fallen back into medieval disarray, to be nibbled to death by the surrounding Hapsburg powers. It was probably best for France that the Magnificent Bastard crushed all his opponents at court, though you couldn't have told them that.[8]

The fact that Étienne Pascal had chosen to live in such fashionable surroundings, so close to the seats of power, suggests that he had more than one reason for moving his children to Paris. Like the cardinal, he was a consummate bureaucrat, a man whose fortunes lay with the administration of the nation. To grow in power and influence would mean to find ways of being close to princes—but, as he would find out, the power of princes can be a force of nature, shifting with the winds of policy. And the man who drove those winds was Cardinal Richelieu.

If Étienne had wished for the life of an intellectual, with an intellectual's reputation, he soon got it. Jean-Baptiste Morin had developed a new technique for calculating longitude, a vital problem for any nation that had dreams of maritime trade, the solution of which meant life or death for sailors. In 1634, Cardinal Richelieu appointed Étienne to a special seven-member committee of scientists and mathematicians called together to evaluate Morin's technique. Richelieu wanted a quick solution to the longitude problem but had little idea of what that involved. After all their long discussions, the committee was not able to declare a solution, though Étienne's gamble had paid off. He had left his hometown for the great city, and the cardinal had favored him—at least for the moment. That would change.

Madame Sainctot's Salon

A gentleman should be able to play the flute, but not too expertly.
—ATTRIBUTED TO ARISTOTLE

A gentleman knows how to play the accordion—but doesn't.
—AL COHN

Fortune favors the brave.
—TERENCE

Père Mersenne's little group was not Étienne's only circle of friends in Paris. Another was the group of pundits, poets, and satirists who gathered around Madame Sainctot, a woman of the court, well known to the queen. Madame Sainctot, like so many court women, was a great beauty with an infamous past. A women of influence, she was a member of the lower nobility who had cultivated the virtues of the courtesan—timing, beauty, cheek, brilliance, joie de vivre, grace, and charm.[9] And she had *l'esprit* aplenty. She was a mistress of the art of conversation, surrounding herself with witty and often outrageous companions, so that her soirees were like her own private court, filled with glittering people who told ribald stories and preached mildly radical ideas. She was also pious in her own way, and sprinkled among the literati was a pinch of clergy. The literati knew their part, as did the clergy, so the conversation was often radical, but not too radical—that could get one a ticket to the Bastille—just radical enough to titillate.

The Sainctot family is little known, though they had their moments. Originating in the Île-de-France region, they were ennobled in 1583, and then confirmed in 1604, likely for bureaucratic service to the king. Jean-Baptiste de Sainctot, the lord of Veymar, had been an ordinary gentleman who had served several terms working in the justice system and as an adviser to Louis XIII. Jean-Baptiste was the uncle of Nicholas de Sainctot, who was the master of ceremonies for Louis XIV. His sister, Madame Françoise de Dreux, was later denounced and tried as a poisoner, though how justly this was done was controversial even then.

Étienne's three children played with Madame Sainctot's children, and so Blaise and his sisters might often have chased one another through the salon until they were caught, or crept about the corners of the adult gatherings, listening in. This would have been Blaise's introduction to the *libertins érudits,* the rising breed of skeptics who would build a new world out of their faith in doubt. Though Blaise was still too young to sense it, the world of adults was changing fast, so fast that a pall of doubt had covered Europe. The old certainties that dominated the Middle Ages had been mortally wounded, but their dying was slow. Blaise himself would become part of the killing. Christendom had been shattered by reform, and throughout most of Blaise's short life, Christians fought one another with a ferocity that only true believers can muster. The old aristocracy, the *noblesse d'épée,* the nobility of the sword, was being replaced step-by-step by the bourgeoisie, that class of new men who spent their lives worrying about money. Blaise's own father, Étienne, was himself one of these new men, as were the Sainctots.

Change was everywhere. Doubt was everywhere. And in times of such uncertainty, some people fling off their clothes and run around proclaiming wild new beliefs and wild new freedoms, while others wall themselves into the fortresses of their beliefs and hunker down. These were the two types who were gathering their forces all across France, first in the Jansenist debate over free will, and then later in the bourgeois revolution that swept aside all their society.

The salon was the gathering place where the intellectually inclined chewed on these uncertainties. There, in Madame Sainctot's salon, Blaise would have heard the wry comments of the satirists and *libertins érudits,* the literary set who first questioned the very basis of the Christian faith and sought to replace it with Deism, with its god of reason, who created the world but who had little to do with the daily affairs of human beings. After a few years, when Blaise entered the society of Père Mersenne, he would have heard the other side of the story—that is, Mersenne's ferocious defense of Christianity and his attacks on the pundits and the scoffers.

Blaise would also have encountered the art of diversion. Within the court, and for those attending it, like Madame Sainctot, one of the most important parts of life was the pursuit of good times. There were, of course, music and the theater, but most of all, there was gambling.

Gambling was everywhere: the aristocracy gambled ferociously; the poor gambled desperately. There was an itch there, an itch that could not be scratched in any other way, an itch caused by the uncertainty of life and a desire for some great event to make it all better. Gamblers place all they are and own at risk, and that is the fun of it. The player throws the dice. The bets are down, the game is committed, and while the dice are tumbling, there is that thrill of mystery and that fear of the future. No one knows what the dice will show, and yet everything rides on it. Whole family fortunes were lost or regained in the tumbling of the dice. In a world where money unrelentingly nibbled at the power of the aristocracy, like hot groundwater eating at stones, what happened in a dice game or card game could mean life or death.

There was something military about gambling among the aristocrats in Pascal's day. Those who were *noblesse d'épée,* ennobled through the sword, or the descendents of long-dead heroes who had risked their lives fighting beside their king, were given lands and titles just as modern armies give medals, and their children and grandchildren inherited their good fortune. These were the old aristocrats, whose families curled back into the dim memory of the nation. Their descendents enjoyed numerous honors and great wealth by the odd accident of their birth. And they assumed that they were superior because of it. But living in the comfortable court as they did, they had few opportunities for heroism, save for the perennial duels that exploded over trifles, so they spent their evenings attending salons looking for a substitute, throwing dice or turning cards.

An ethical stoicism hid behind this martial understanding of gambling. In classical culture, the soldier, and therefore the noble, encountered a choice when confronted by chance events. If a battle turned against you, you were expected to stand up to your fate and take it without flinching. If a battle went for you, you were expected to do the same. When Fortune turned against you, you could either buckle before fate and cry to heaven over your bad luck, or you could stand tall and accept whatever came, the good and the bad, with equal aplomb. This was *areté* for the Greeks and *virtus* for the Romans, an ethical quality of soul that raised the aristocrat above the common herd.[10]

Dicing in one form or another is the oldest of all gambling games. Card games became popular only after the invention of printing, but throwing the astragali is as old as civilization. Egyptian inscriptions have depicted dicing games from 2000 B.C.; the Chinese have references dating back to 400 B.C. The word *dice* is the plural of *die,* which comes from street Latin *data,* which comes from *dare,* "to give." To throw the dice is to face that which is given by the gods, by powers higher than human. It is to face reality at its most mysterious, like standing unflinching before the thunderstorm.[11] The tradition, coming down to us from the Romans, was that there was a goddess, Fortuna—or, in Greek, Tyche—who ruled each turn of the card, each toss of the dice:

> *O Fortune*
> *like the moon*
> *changeable in state*
> *always waxing*
> *and waning;*
> *detestable life*
> *first oppresses*
> *then assuages*
> *as its whim takes it,*
> *poverty*

and power
it melts them like ice.[12]

This was the dominant belief, the common sense throughout the Middle Ages and the Renaissance. The tavern was the goddess's place, the place where the courageous wrestled with her nightly. It was a place apart from the rest of the world, where the ordinary rules of society did not apply. Those who forgot the slipperiness of Fortune were doomed to suffer under her wheel, for such fools were guilty of hubris, vaunting pride, the source of all sin. Thus, in its earliest incarnation, gambling was a warriors' game, a sparring match between the knight and the goddess. Like jousting, it was both a sport and a reenactment of the truth of life. The first Christian dice game, *le hasard,* was brought back to France by crusaders, and it offered the knights a chance of doing battle in the drawing room, a way of civilizing war itself or of bringing war home with them. William of Tyre tells the story, in his *Historia rerum in partibus transmarinis gestarum,* about a crusader army forced to stay in a Syrian castle named Hasart. Far from battle, the men were so bored that they invented a dice game, *hasart,* and enjoyed it so much that when they returned to Christian lands, they brought the game with them, and that game gradually mutated into the modern game of craps. The goddess, meanwhile, the personification of surprise and of caprice, handed out good and evil as she desired.

Young Blaise likely attended one of Madame Sainctot's soirees and peeked into the room where the adults were playing adult games. As a child, he would not have understood the subtleties. Gambling was a demonstration of the aristocrats' superiority over the bourgeoisie, who spent their lives intertwined in the affairs of money. It was therefore winked at, though it was illegal. The fact that the aristocracy could practice it with impunity was another sign of their inherent superiority. Meanwhile, the bourgeoisie looked at gambling as a moral evil, as a languid vice of the nobility, whom they saw as living misspent lives. Not that they didn't yearn for the aristocratic pleasures themselves. After the French Revolution, the bourgeoisie happily picked up all the vices that the aristocracy had dropped on the way to the guillotine. Before the Revolution, however, their condemnation of gambling was largely a product of their own fear of falling. Because their social standing depended so thoroughly on money—to accumulate it, to use it, and to disperse it—any cavalier use of money, especially in the manner practiced by the aristocracy, was treated almost as a personal affront. Nevertheless, as a courtesan, Madame Sainctot would have practiced those very vices, and allowed them to be practiced. How else could she have looked like an aristocrat herself?

Oddly enough, the bourgeoisie were caught between the martial games of the upper classes, partly expressed through high-stakes gambling, and the endemic gambling practiced by the poor. Whereas the aristocracy could disdain money because their social position did not depend upon it, the poor had no money, or too little of it, and so did not fear falling from their social position, because they had already fallen. It was the unique and exquisite plight of the middle classes, especially those who had been recently ennobled or were on their way to becoming noble, that no matter how high they climbed, their lives were still dependent upon money, and they still suffered the lightly veiled contempt of the aristocracy, whose social position was so much more stable than their own. When they gambled, which they did regularly in order to fit in with the court, they did so with fear and trembling—which, of course, they could not let show without also displaying their middle-class origins.

Thus the Pascals, who lived on a fixed income, settled into a life at court, where people nightly put their children's future at risk, and pretended to be amused. Blaise, ever observant, watched the great show, and learned more than anyone expected.

Le Libertin Érudit

The sworn enemy of impiety,
The Deist lives in peace with all people,
The only true observer of religion
He worships the Author of the earth and sea.

—QUATRAINS DU DÉISTE

Pascal was fourteen years old when he entered Père Mersenne's seminar, and sixteen years old when he presented his first paper, his one-page essay on conic sections. He was no doubt the enfant terrible, the prodigy, the boy from whom everyone expected great things. At this point in his life, young Blaise was little more than an extension of his father, Étienne. Always under his father's thumb, he showed little evidence of adolescent rebellion, for nearly everything he did was to gain his father's approval. For some in the group, especially Descartes, he may have seemed too proud, too confident for a boy his age, but then again, how would any sixteen-year-old react to praise lathered on not only by his father but by his father's friends?

One can only imagine the conversations held in that room: the reports from Père Mersenne about the latest scientific discoveries; the terrifying news of Galileo's trial, which everyone in the group—Catholics all—deplored; Étienne's discussions about the longitude question; Père Mersenne's latest attempts to predict prime numbers; and the most recent controversy of several on Descartes' algebraic geometry and on his philosophy. In the back of everyone's mind, however—troubling them all, including Descartes—was the secretive group of doubters and free-thinkers who had burrowed their way into the intellectual life of Paris. These were the *libertins érudits,* that collection of atheists and material-ists, like Cyrano de Bergerac, whose ideas were being quietly discussed in the salons and universities around the city. Père Mersenne wanted to refute them. Descartes wanted to employ their very doubt against them, to make it impossible to doubt by finding one undoubtable truth.

The roots of unbelief go deep in Western culture, predating Christian-ity, back to Rome, back to Greece, perhaps as far back as religion itself. Atheism and materialism form a tradition of their own, a belief system of their own, a vision of life with its own body of arguments, a tradition that is forever the negative of religion, its dialogic partner in the great conver-sation. Who knows why some people are religious and others are not?

For Pascal, in his later work, those without faith were predestined by God to be faithless. He believed that God chose to be revealed to some people and to be hidden from others. This was his theory of the *Deus absconditus,* the hiding God. The issue is still alive. For contemporary bio-chemists, there is talk of a God gene—a gene that, expressed in one way, leads to spirituality and an abiding sense of the presence of God and that, expressed in another way, leads to materialism and the sense that human beings are but bags of chemicals. Perhaps these two points of view are not that different, in that they both assume a manner of fate. However, some hear the music of the angels, and some do not. The Gospels often point to the same experience: "Those who have ears to hear, let them hear!"[13] Why this is so is the strangest of things, for luck and grace seem like the twin sides of the god Janus.

Everyone believes, for there is no other way to live. There is no way out of this. Pascal saw this in his famous passage on the wager. Even those who say they *know,* that they have no need of belief, are throwing the dice. They are just throwing them harder than most.

The doubters in Pascal's time were largely Deists, those who rejected the idea of Providence, the belief that God is directly involved in the lives of people. This new religion was a rejection of the heart of Christianity and not a mere tinkering with the theological details, as had occurred in the rift between Protestants and Catholics. Moreover, it was a perfect religion for the growing scientific mentality of middle-class capitalism, with its God who, once having created, stood back and observed the working of the world without getting his fingers dirty, without trying to alleviate pain, or save souls, or punish the wicked. That left room enough for human enterprise to do the rest.

Deist manifestos, often written as poems and often anonymously, circulated around Paris. The most famous of these was the *Quatrains du déiste,* which rejected all anthropomorphic images of God as superstition. How can the Eternal God be like mere human animals? To be Eternal, God must be something else, something Other, something abstract, like Plato's Ideas:

> *Since the Eternal Being eternally*
> *Knows only great beatitude, perfect and all sufficient, . . .*
> *Is not the one sunk in superstition insane*
> *To imagine Him both unchanging and changeable*
> *Inflamed with vengeance and offended by a thing of little consequence*
> *An enemy of tyrants, yet more redoubtable than they?*
> *And is ["le superstitieux"] not yet again insane to imagine [God],*
> *The Sovereign guide of the whole universe*
> *And at the same time believe that He lets himself be swayed*
> *According to the passions and human nature?*[14]

For the Deists, God was beyond humanity, and that meant that such human notions as God's wrath, God's love, and God's law were all foolish superstition. Père Mersenne was furious with this last part. In his *Impiety of the Deists,* he attacked the *Quatrains* as the summation of every doubting impiety since the beginning of Western culture: "I think that your poet has assembled all the impieties of Lucian, of Machiavelli and of all the libertines and atheists who ever existed . . . to argue that Divine Law is but an imposture."[15] Mersenne took the whole business to be a plot to ensnare the foolish and naive in this world, to draw them away from the faith. Later Pascal, in one of his most Jansenist moments, took up

Mersenne's banner and carried it further, claiming that the Deists were little more than atheists.

But what Mersenne and Pascal did not realize was that Deism had been given room to grow in Europe because of their own work. A few years after his sojourn with Mersenne's seminar, Pascal proved the existence of the vacuum—that there could be empty space in the world, space not filled with anything. And behind this concept was the new Copernican universe that took humanity out of the center of things and sent the Earth spinning around a mediocre star in a nondescript corner of a medium-sized galaxy. How could the creator of all *that* be born to a single species of animal living on a lukewarm nothing planet like Earth?

This was the new universe that they were living in, the universe that was only beginning to reveal itself, the universe that gave Pascal such terrors in his last days. And the Deists had a better grip on it than the Christians. It was not until the twentieth century, when this universe proved even stranger than the Deists could imagine, that the increasingly strange vision of a God who becomes human could make sense once again, and even then only through the back door. With a universe as odd as this, anything can happen.

Charming the Cardinal

*Grown-ups never understand anything for themselves, and
it is tiresome for children to be always and forever explaining
things to them.*

—ANTOINE DE SAINT-EXUPÉRY, *The Little Prince*

D isaster! In March 1638, Étienne Pascal fled Paris, running
for his life. Two of his friends had been put into the Bastille, while those who had evaded the cardinal's agents were looking for
some hole to crawl into. Étienne ended up back in Clermont, in his hometown, where the network of aunts, uncles, cousins, nieces, and nephews
could hide him better than he could hide himself in Paris.

As with most things, this drama began with the cardinal. When
Étienne had brought his children to Paris, he had several financial assets:
his house in Clermont, on the rue des Gras, and his office in the Cour
des Aides—an office, like many others in France at the time, that could
be bought and sold. When Étienne sold his office, he took most of his
money and invested it in French government bonds, or *rentes*. These paid
the interest of 1 livre per year for every 18 livres invested at the time of
his investment. This made Étienne's investment worth 65,665 livres. However, within a few short years Cardinal Richelieu brought France into the
Thirty Years' War on the side of the Protestants because he feared the

spread of Hapsburg power. We must remember that the Hapsburg family ruled Austria, parts of Germany, what is now the Czech Republic, a good chunk of Italy, and the entire Spanish empire. The Hapsburgs were also ultra-Catholic and were using their power to fight the increasing influence of Protestantism. Richelieu, a cardinal of the Roman Catholic Church, worried more about infringements on French sovereignty than about the life and health of his church, and so he declared war on the Holy Roman Emperor. The problem with war, however, is that it costs a lot of money, and the cardinal had to get that money from somewhere. The cost of the war had nearly bankrupted the nation, and so Richelieu, to solve his financial problems, decided to default on his government bonds. The value of Étienne's investment dropped from 65,665 livres to less than 7,296.

Money makes the world go round, and this was also true in the seventeenth century. The aristocracy had their privileges, granted to them by birth and family history; the poor had almost nothing; the rising bourgeoisie had money, and money was slippery. The entire story of the events leading up to the French Revolution can be summed up in this truth. The seventeenth century was not merely the time in which science began to take hold, but also the time in which money began to take hold, to affect even the aristocracy in their complacent privileges. But it was the middle class, then as now, who always felt the pinch, who, unless they secured their money, could easily fall back into the nameless masses of the poor. Understandably, then, Étienne and the other investors grew disturbed over the cardinal's decision. That March, in 1638, Étienne Pascal, the man of good breeding, the man of science and mathematics, the enlightened teacher, the socialite and intellectual, joined the other investors in a protest, which, as protests will, got out of hand. Someone made threats, and someone acted violently, and someone was openly seditious. The cardinal, in a snit, responded in kind and ordered his agents to gather them all up and throw them in prison. If Richelieu knew anything, it was how to put down protesters.

In full flight, Étienne left his children behind to be cared for by friends and by his domestic servants. Imagine the fear of his children: they had lost their mother when they were only babies, and now their father, their only protector, had been forced to run for his life, or at least for his freedom. But the Pascals were fortunate in their friends, especially Madame Sainctot, who gathered the children around her and began to scheme for Étienne's return and rehabilitation. The friends had one hold card. Blaise was not the only talented Pascal; his younger sister, Jacqueline, was an accomplished poet, and had been even as a child. She was a pretty child and, like Blaise, intelligent beyond her years, the kind of child who makes adults coo when they are anywhere nearby. She was also a sunny child, happy most of the time, at least according to her sister, Gilberte—though Gilberte cannot always be trusted, because she tended to idealize her family. But like her brother, Blaise, Jacqueline was also stubborn.

Her father had once assigned Gilberte the job of teaching Jacqueline to read, and for some time the little seven-year-old seemed impervious to her sister's instructions. Then, by accident, Gilberte read Jacqueline from a book of verse, and it changed everything, for the little girl seemed captivated by the music and rhythm of the language. From that point on, it was easy. Four years later, along with the two daughters of Madame Sainctot, Jacqueline wrote and performed in a five-act play, written entirely in verse. The play was the talk of the fashionable ladies of Paris, and the sudden notoriety started the young Pascal on a literary career of her own.

In 1638, the king's wife, Anne of Austria, conceived after twenty years of waiting, if not twenty years of trying. Her husband, Louis XIII, was a scrupulous man, and cold to nearly everyone, especially his wife. As was all too common in kings at that time, he didn't like his wife very much. She had been foisted on him as a political choice, not because of any compatibility, but rather as a way of connecting the French throne to the all-powerful Hapsburgs. There is some doubt that he even consummated the marriage. No one would admit that, however, because it would have led to war. The people of Paris had begun to wonder if the king and queen would ever conceive. Mostly, they were simply frustrated with the royal pair: would they ever get around to doing their duty? At a time when producing a legitimate heir to the throne was the principal duty of any king, this was an important question. No one had forgotten the endless dynastic wars that had beggared Europe throughout the Middle Ages.

But then, like Sarah the wife of Abraham, Anne came up pregnant just when the people were beginning to lose hope. This child would live to become Louis XIV, the Sun King, the man whose lavish lifestyle would set France solidly on the path of revolution. The court gathered around Anne to pamper her in her delicate condition and schemed to find diversions to keep her entertained. Unlike Louis, Anne was a party girl and had the reputation of someone who liked to have a good time. Just before he went fugitive, Étienne and his friends brought Jacqueline to court to read some of her sweet poetry in honor of the queen and her pregnancy. As expected, the little girl charmed both the king and queen, and both praised her and fawned over her, no doubt to the delight of her father.

This was the arrow that Étienne's friends used against the cardinal. Sadly, soon after her triumph with the king and queen, Jacqueline contracted smallpox, a nightmare common enough in Europe throughout the seventeenth century, one that we have nearly forgotten in our time. Hanging near death for days, she struggled against the virus, while her father sat beside her bed. Slowly, she recovered, though her perfect face was marred by the scarring the disease left in its wake. Jacqueline later claimed that it was this loss of her beauty that was God's greatest gift, for he took her off the worldly path that she had been set upon and set her on a new road that led more directly to God.

Then came the terrible month of March and the reversal of Étienne's finances, the ill-conceived protest, and Étienne's flight from the city. Étienne's friends quickly wove the three children into their schemes. Jacqueline the charmer would be introduced to the cardinal and read a few poems to him that had been written in praise of him and his administration. The other children would look on as cherubs. It was cleverly done, for the cardinal was likely to be undone by the little poetess whose charms had only been added to by the tragedy of her illness. Could the cardinal not also see her as an orphan, now that her father had been forced to flee the city? Moreover, the cardinal had made a great enemy out of Anne of Austria, whom he never really trusted, since he reckoned that her loyalty was more strongly directed toward her brother the king of Spain than it was to her husband, the king of France. Richelieu had a cruel tongue and had made Anne squirm under it more than once, and had done so in public. So how could Cardinal Richelieu allow his enemy the queen to show more compassion for the sweet Pascal girl than he did? And of course there were plenty of courtiers present to remind him of just that. Jacqueline played her part to the hilt; after reading her poems, she sat upon the cardinal's lap, where he kissed her cheek and praised her over and over, calling her a sweet child and a wonderful poetess. It was just at this moment that Jacqueline leaned against the cardinal's breast and asked a favor of him. *Please bring my father home,* she said. *He is most terribly repentant. Bring him home and you will see what a good servant you will have in him.* The rest of the court smiled on benignly, including some of the king's and queen's closest friends. The cardinal was caught like a fish. What else could he do?

Conic Sections

There is no royal road to geometry.
—EUCLID TO PTOLEMY

J acqueline's gambit had worked. According to her sister, Gilberte, "M. the Cardinal said to her: 'Not only do I grant your request, but I heartily desire its fulfillment. Tell your father to come to see me in all confidence, and when he comes he should bring his whole family with him.'"[16] Jacqueline begged her father to return home and present himself to His Eminence, which, in spite of the cardinal's kind words, was not a completely safe thing to do. What the cardinal might say to a sweet little child, so lately the favorite of the king and queen, was one thing, and what he might say to the father, so lately the cause of civil unrest, was another. But as it turned out, the word of Richelieu was good, at least this time. Étienne Pascal appeared before him to make his apologies and to beg forgiveness, and nothing particularly dangerous happened to him. Jacqueline must have been quite the hit, even with the inventor of realpolitik, for not only was her father forgiven, but six months later he found himself with an appointment as the king's commissioner of taxes in the city of Rouen.

Rouen is a seaport on the river Seine in Normandy, slightly northeast of the American beachheads of World War II, and is mentioned sooner or later in nearly every movie about D-day. It was built as close to the mouth of the river as possible and still spans the river with bridges. In its youth, Rouen was a trading center for Celtic merchants; later, it became a Roman outpost, and later still a base camp for Vikings. In the nineteenth century, its already famous cathedral was sketched and painted over and over by Monet, to capture the subtle moods of the light. It was a thriving seaport in Pascal's day as well, and trade brings taxes, which bring tax collectors.

In spite of all his outward kindness, Cardinal Richelieu never completely forgave the Pascals, or anyone who opposed him. What he had given to Étienne was both a reward and a punishment, for the region was aflame with a vicious tax revolt. Royal commissioners for taxes were roundly hated by the people, and France's involvement in the Thirty Years' War had been squeezing them into revolt or starvation. There were taxes for just about everything, and fees on the taxes. As usual, the rich understood the system and used it for their gain. The king's creditors received the privilege of levying fees upon the people in a particular region, thus repaying the debt without much bother to the king, much as if Chase Manhattan Bank held a promissory note from the president, and in payment received the right to wring extra money out of the people of Minnesota. This saved the king the bother of trying to find the money to pay back his creditors, and if in the process he added one more tax onto the already overburdened people, what did that matter? Understandably, civil uprisings were popping up all over France, explosions of peasant and minor bourgeois fury that would brood throughout the reign of Louis XIII, then into the reigns of Louis XIV, Louis XV, and Louis XVI, and suddenly flame into the Great Revolution. But the French kings did not know that they were signing away their future just to unravel the crisis of the moment, for the easy way to solve problems is rarely the best way. In 1639, the year that Étienne Pascal brought his children to Rouen, uprisings burned across the entire region, exploding overnight and then burning on sometimes for weeks, sometimes for months.

Such events occurred all over France. It is likely that nearly every day young Blaise overheard a report of an uprising somewhere—a riot, a brawl, a murder. Years later, in the *Pensées,* he wrote about kings and their use of power, about the pain they caused, about the death. Perhaps his own sympathies during those years may eventually have come to rest with the people, the common folk.

This was a creative time for Blaise, a time when he started to build his reputation. The jury was out on his personality, however. Some said that young Blaise Pascal was a prodigy, others that he was an arrogant boy. Probably both views had some truth. Certainly, he was the overprotected son of a rich father, a father who had achieved intellectual fame and who wanted his son to do the same. He was a driven boy, and yet one can forgive much of his edge because of his lust for knowledge. He was manically curious, with an encyclopedic, roving mind that attached itself to one mathematical problem after another, a lamprey chewing through the skin of a shark until he had penetrated the problem deep to the bone. What was best about him was his openness to unfamiliar ideas and his willingness to accept the evidence of his eyes. These gifts would serve him well in the years to come.

Gilberte was then twenty years old and Blaise nearly seventeen. Rouen was an old city, with winding, narrow streets and tall half-beam houses. The Pascal house was in a compact neighborhood near the monastery of Saint-Ouen, what would have passed for a suburb in that time, on the northwest edge of the city, an area mainly occupied by bureaucrats and their families. Their friends and neighbors were mostly government employees at one level or another. Churches were all around them, with shops sprinkled here and there. Nearby was the rue du Gros-Horloge, where St. Joan of Arc had been burned as a heretic. The gothic spires of the church of Saint-Ouen dominated the skyline, with the building's stained-glass windows depicting biblical stories and moments in the lives of the saints. When the sun shone through them, the interior of the church was cast with rose- and gold-colored light. In the summertime, flowers peered blue and red and white above the lips of planter boxes, while off in the distance the sea boiled up hillocks of vapor that marched onshore as the day waned.

Blaise was now in the coils of adolescence. He was not a handsome boy, small for his age, thin and frail looking, snappish one moment, sentimental and pious the next, slouching between insecurity of body and arrogance of mind. His sister Gilberte had maintained her dark beauty, dark eyes, dark hair, white skin, and elegant figure, and she was surrounded by young men. Jacqueline, though her face had been spoiled by smallpox, was lovely enough to have her share of suitors. Étienne tried to find a husband for her several times, but it never quite happened. Jacqueline showed little interest. Outwardly, however, the Pascals were the perfect provincial family—middle-class, comfortable, acceptable, with hints of great expectation sprinkled through. Inwardly, they were hyperintelligent, given to emotional extremes, and not a little eccentric. Though Blaise had shown the first signs of great promise, it was Jacqueline who kept winning prizes for her poetry, and it was this rather than romance that consumed her life. Blaise had yearned for celebrity since childhood, and Jacqueline's early success must have galled him in his private room, though he would have smiled and applauded her in public. Moreover, while Blaise was utterly under his father's thumb, following out his father's program, Jacqueline showed an alarming independence, something that was simply not done among their kind.

While in Rouen, the family had met Pierre Corneille, the Norman tragedian, who quickly befriended them and encouraged Jacqueline to pursue the life of a poet and playwright. It seemed at first that she was on the verge of doing so. Meanwhile, Blaise received a copy of Desargues' book on conic sections and, egged on by his father and by Mersenne, set out to make his own contributions. Living in Rouen at the time, he was off in the provinces, like Fermat in Toulouse, and had to rely on letters and gossip for news of the intellectual tides of Mersenne's academy. But the monk, the great communicator, kept everyone apprised of the latest, and made sure that the right books ended up in the right hands.

While Blaise scribbled his notes on conic sections, crowds gathered in the street below and shouted angry insults at the government employees in the neighborhood. Richelieu's gluttonous taxes, the very thing that the Pascal family had come to Rouen to administer, had crushed the common people into rebellion. Brush fires of plague had also erupted in parts of the city and throughout Normandy. In the summer of 1639, just as the

Pascal family had taken up residence, the city exploded with riots. Gangs of looters roamed the streets, singing bawdy songs and throwing curses. For the first year of the Pascal family's stay in Rouen, they must have felt as if the people's vengeance would swallow them whole.

Finally, Blaise sent his first work of serious mathematics off to Père Mersenne. It was a short piece, a pamphlet entitled *Essai pour les coniques*, or *Essay on Conics*. In it, Blaise outlined his proof for what has been called the Mystic Hexagram. Here is a short description of the concept:

1. Take a cone:

2. Take a simple plane, and slice the cone in two.

3. If the plane is straight across, the section cut out will be a circle. This is the specialized case:

4. If the plane is at an angle, the section cut out will be an ellipse. This is the more general case, because ellipses can be squat or long, thin or nearly round:

Because Pascal wanted to prove a general theorem, he took the case of an ellipse:

5. Draw a hexagram, a six-sided figure, inside the ellipse. The hexagram

6. Now, take a pencil and make big dots on the vertices of the hexagram, and draw lines

does not have to be regular.

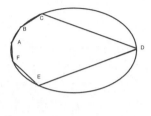

between the vertices. Then, extend the lines out to where they cross.

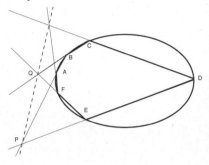

7. The three points of intersection where the lines cross will always form a straight line, for any conic section and any hexagram.

Mersenne was deeply impressed when he received a copy of Blaise's pamphlet. Everyday geniuses like Descartes are one thing, but a child prodigy is another. There is something divine about a child, even a teenager, who shows promise beyond his years. Blaise was thirteen years old when he first began attending Mersenne's seminar, and everyone thought that he showed great promise. But unfulfilled promise is only air until the prodigy actually does something, and Blaise's pamphlet was not only the first sign of that fulfillment but also the vindication that Mersenne had been waiting for. He had been telling everyone about the bright young son of Étienne Pascal for years, and if that boy had done nothing with his talent, it would have been sad for Blaise and embarrassing for Père Mersenne.

At once, Mersenne sent word about the younger Pascal's success to his contacts throughout Europe. Scientists and intellectuals as far away as Poland heard about the young geometer. Meanwhile, Descartes, too, heard about the pamphlet, and on a visit to Mersenne, the monk showed him a copy. Descartes, who was then forty-four and still angry about Étienne's criticism of his own geometry, was not impressed, and instead grumped like a grandfather with a bad tooth. "I do not find it strange that he has offered demonstrations about conics more appropriate than those of the ancients," he said, "but other matters related to this subject can be proposed that would scarcely occur to a sixteen-year-old child."

The Arithmetic Machine

I propose to consider the question—Can machines think?

—ALAN TURING

B y 1642, the Pascals were well settled into the local scene in Rouen. That year, Gilberte, living in Clermont, gave birth to a son, Étienne, named for her father. Also during that year, the fearsome Cardinal Richelieu succumbed to his many diseases and died. Within a year, his master, Louis XIII, would follow. The English civil war between the Cavaliers and Roundheads began in 1642, and Galileo Galilei died of illness and old age, still under house arrest. Closer to home, the violence of the Pascal family's first days in Normandy had subsided, and Étienne had begun his work in earnest as the chief tax collector of the region. His biggest problems had become mathematical. With the uprising, the registers had been destroyed, the office was a mess, the state of collections was disorganized, and Étienne had to spend long, laborious hours calculating, calculating, and calculating. He worked into the night every night, and soon the exhaustion that hung about this new honor was like a funeral wreath. Étienne grew snappish and complained about his health. "For four months now, I have not gone to bed more than six times before two o'clock in the morning," he wrote to Gilberte.

Blaise, then barely nineteen, volunteered to pitch in, but he was quickly buried by the mountain of work, and since his health was ephemeral, he found himself in some distress. His head hurt. The problem was complicated by the complexity of the French currency. One French livre equaled 20 sols; one sol equaled 12 deniers, so that the steps up and down the currency were uneven and therefore not easily tabulated. One livre was therefore 240 deniers—which was unnecessarily complex. Each set of calculations, therefore, was exquisitely tedious, just the kind of work that sucks the life out of one's soul. There had to be an easier way.

Out of pure necessity, Blaise cast about for a solution, and found it in a machine. Blaise saw fairly quickly that the kinds of calculations that his father's work required were mechanical and could be done by a machine. But what kind of machine? The machine he came up with, the Pascaline, was the first calculating machine in the modern style. The centuries have refined the mechanism, but the basic concept underpinning the logic of Pascal's device and the logic of an electronic computer are not that different. In essence, Pascal set up a series of gears that moved one way— forward but not backward. Inside of a long narrow box, he fixed eight cylinders, with the numbers from zero to nine printed on each one. Each cylinder had two sets of numbers, however, one descending, at the top of the cylinder, and the other ascending, at the bottom. Therefore, the top numbers were used for subtraction, while the bottom numbers were used for addition. The user would choose one operation over the other by sliding a metal bar over the unused portion of the cylinder, to highlight the calculation at hand. Underneath the row of cylinders, set horizontally across the top of the box, was a sheet of metal that hung vertically, perpendicular to the row of cylinders. On that sheet were eight wheels that turned. The wheels were attached to a set of gears inside the box with various numbers of teeth on them. In order to account for the complexity of the currency, some of the wheels had ten spokes on them, others twelve, and others twenty. A series of gears connected a series of wheels on the box lid to the drums. To calculate, the user took a metal hook and turned the wheels on the lid. In order to carry numbers past ten over to the next column, Pascal set up an escapement arm that regulated the motion of the wheel that stood for the power of ten.

Thus he mechanized addition and subtraction, but he failed to get his machine to multiply and divide. There was a calculating clock that had been invented in Germany back in 1624 that could do this, but it worked entirely differently. The problem was finally solved by Leibniz, who tricked up the Pascaline by adding stepped teeth that would repeat adding and subtracting over and over again to simulate multiplication and division.[17]

So the machine worked, and it took on much of Étienne's burden. But Pascal the capitalist knew that he had something worth marketing, something that could make him famous throughout Europe, something beyond the work of a "clever child," as Descartes had called him. But how to get the thing mass-produced? There were plenty of master craftsmen in Rouen. The city made some of the best glass in Europe, and there were metalsmiths and watchmakers aplenty, but none of these men understood Pascal's science, and Pascal did not understand their craft. He had the devil's own time of getting the workmen to produce what was in his head. In his dedicatory letter to Chancellor Séguier, he complained about his difficulties: "Knowledge of geometry, physics, and mechanics furnished me the design for it, and assured me that the employment of it would be infallible if some workmen could make the instrument whose model I had conceived. But it was at this point that I encountered difficulties as great as those I wanted to avoid, and to which I was seeking a remedy."[18]

The difficulties he encountered exhausted him, and his health withered. More weakness, more headaches, more pains in his legs. Still, many people were encouraging him to continue the work—Père Mersenne, of course; Chancellor Séguier; his father; the prince de Condé; and others—and so he pressed on, revision after revision, fifty revisions, until the thing was ready for market. During this process, however, a local watchmaker, obviously in league with some of Pascal's craftsmen, came out with an inferior version of the Pascaline, and claimed credit for it. Pascal had been pirated before he even finished the work. Furious, he cancelled the entire project and fired his employees. But then his friends kept on, and even more letters of encouragement from powerful people arrived in Rouen, some with advice on dealing with craftsmen, some with advice on marketing the machine once it was done. Encouraged, Pascal worked on into 1644, when he presented the prototype Pascaline to the chancellor, along with his official letter of presentation. He included a notice along with the letter, warning people of pirated editions of his machine that were floating around and asking them to avoid them.

The one problem that Pascal had with his machine was the price. As is the case with almost all truly new inventions, the cost of producing Pascalines made them rare.[19] Too often, inventions that have the potential to change the world do so only after many years. At first, they remain toys for the very rich, toys that such people buy for their obvious potential rather than for their present use. The local butcher and baker still had to count on their fingers. Nevertheless, the prince de Condé bought one, as did the queen of Poland and Louise-Marie de Gonzague. Wladyslaw VII also bought one, and asked if it could be built to accommodate Polish money.[20]

Even if there was little profit in it, the Pascaline was a hit. Roberval made demonstrations of the device all around Paris, and Mersenne celebrated his young protégé's genius to all his correspondents. Pascal was suddenly the talk of the salons, the boy genius who lived up to his great promise. Throughout all of this, however, Descartes kept a grumpy silence. Sooner or later, he and the boy would go to war; he was certain of it.

The Jansenists

*The weakness of little children's limbs is innocent,
not their souls.*

—ST. AUGUSTINE, *Confessions*

*Two cities have been formed by two loves; the earthly by
the love of self, even to the contempt of God; the heavenly
by the love of God, even to the contempt of self.*

—ST. AUGUSTINE, *De civitate dei*

L ate in 1638, the same year that Étienne Pascal went into hiding from the cardinal's agents, a prisoner languished in one
of Richelieu's dungeons. He was not an ordinary political prisoner, but a
priest, an abbot, and one of the great religious leaders of the day. He was
Jean Duvergier de Hauranne, the abbé de Saint-Cyran, one of the founders of a spiritual movement within Catholicism called Jansenism, a new
form of strict Augustinianism—a Catholic version of Calvinism, at least
according to the Jesuits. In happier times, he had been celebrated across
France, and his name had penetrated into every corner of Europe. The
destinies of these two men, one in hiding and one in prison, would one
day cross. Étienne's children would become the abbot's followers, to their
joy and to their sorrow.

Saint-Cyran's movement would gather a number of great lights to it-self, men like Antoine Arnauld, Jean Racine, and, of course, Blaise Pascal. Saint-Cyran also insinuated himself into several influential convents, most notably Port-Royal des Champs, an old Cistercian monastery turned fashionable center for devout young ladies, including Blaise Pascal's sister Jacqueline. He became a strange attractor for devout young men like Blaise, who wanted to live an odd variant of monastic life, and a center for a highly ascetic and highly nationalistic brand of spirituality based on a strict interpretation of Augustine's doctrine of original sin, according to which the vast majority of humankind, including most Christians, was doomed to hellfire. It was all fairly depressing.

France was undergoing a major Catholic revival in the seventeenth century. There were two lines of thought generating it, turning on the question of whether human beings were truly free and whether human nature was capable of goodness after the Fall. Each of these two lines of thought had a political dimension as well. The first was a Gallican line that wanted to promote the local French church as a separate cul-ture distinct from Roman oversight. Included in this camp was Pierre de Bérulle, the founder of the Oratorian order, and later Cornelis Jansen and the abbé de Saint-Cyran, the founders of the Jansenist movement. Characteristic of this group was a deep pessimism about human na-ture—a belief that, left on their own, human beings were incapable of choosing the good and that human nature needed to be disciplined to the point of annihilation. This was straight Augustine, and those who followed this line of thought were proud to trumpet that fact. Bérulle, who was in many ways the spiritual father of the entire French revival, was close to a number of the great spiritual leaders of the day—St. Vincent de Paul, St. Francis de Sales, and St. Jeanne-Françoise de Chan-tal. His own spiritual thinking, which was Neoplatonic in origin, called for the annihilation of all of the natural powers of the individual and their replacement with a strict adherence to Jesus in the Eucharist. The purpose of life for Bérulle could be found in the conforming of the individual to God's will, which would require self-immolation, burning out all instances of self-love.

Bérulle likened this to a Copernican revolution:

An excellent mind of this century wishes to hold that the sun is at the center of the world and not the earth; that it is immovable, and that the earth, in proportion to its round shape moves in reference to the sun. . . . This new opinion, scarcely followed in the science of the stars, is useful and ought to be followed in the science of salvation. Because Jesus is the sun, immovable in his grandeur and moving all things . . . Jesus is the true center of the world and the world ought to be in continual movement toward him.[21]

One of Bérulle's disciples was the man who would eventually become the prisoner at Vincennes, the abbé de Saint-Cyran, who took Bérulle's spiritual teaching and, with the help of his friend Cornelis Jansen, ran with it, formulating a strict construction of the Augustinian theory of grace and free will. It was their extremism that cast light on the other line of thought during the revival, which was spearheaded by Francis de Sales and the Jesuits. Francis de Sales eventually wrote his greatest work, *An Introduction to a Devout Life,* with the idea that people had a natural inclination toward God—that concupiscence, the propensity for evil, was not the only player in the human soul, but that the primary movement of the human heart was for its Creator. This mirrored the Jesuit idea, which came directly out of the personal experience of St. Ignatius of Loyola that every human being had the possibility, the duty even, to make a choice for God or against God—to march under the standard of God or under the standard of the devil. Therefore, the Jesuits opposed Augustine's limitations on human freedom. And their liberality, I would argue, was the wellspring from which the modern idea of liberty flowed.

For Saint-Cyran, these milquetoast spiritualities were just too soft on sin. A Christian should be made of sterner stuff. His official portrait shows a man dressed in cassock and surplice, nearly bald and with a full beard, his lips pressed tightly, his eyes sliding away, as if with an irreparable sadness. The painting does not do him justice, however, for he was one of the most successful religious salesmen of the century. Vincent de Paul and Francis de Sales had both fallen under his influence for a time, but after suffering through painful crises of faith that involved serious anxiety over whether they were included in the elect, they both broke with him, saying that they were horrified by his religious paranoia. Nevertheless, he was able to convince others—those more attuned to his brand of religion, even powerful women like Mère Angélique, the abbess of Port-Royal, the dowager empress of religious women in France, and others. He taught them that only a life of extreme self-punishment could lead to salvation, an annihilation of all attachments and worldly associations, an extreme discipline of the body, to the point that the nuns of Port-Royal encouraged the growth of body lice and other vermin as an act of mortification.

Richelieu had thrown Saint-Cyran into prison on the exaggerated charge that he had criticized the king's prayer to the Blessed Virgin, where he asked for her special protection for France. This was rude at best and unpatriotic at worst. But this charge was only a convenience, for Richelieu had been cultivating his disappointment with Saint-Cyran for some time. They had once been acquaintances, and had met regularly to discuss things theological. But Richelieu did not care for the penitential way of life, and he eventually broke with the entire movement and sided with the Jesuits, who were the liberal modernists of the time, for they were men who understood politics and did not shy from engagement with the world.

Nor did it help that Saint-Cyran had inadvertently become tangled in one of the king's fits of conscience. Absolute monarchs of the period lived in a strange world. On the one hand, from the time they were children they were told that God himself had blessed them with royalty and that the exercise of power was not only their privilege but their responsibility. The king had to rule, and by that very fact he lived in an alternate moral universe. Not even his spiritual director complained too much about his lovers, which, in the case of Louis XIII, were generally older men; nevertheless, his conscience, if he had one, could be a source of political leverage. Unfortunately, Louis XIII had a conscience, and a touchy one at that. He was an extraordinarily complicated man, who could ruthlessly order the storming of a castle or the execution of an enemy one day and then wring his hands over his failure to achieve perfect contrition the next.

Louis XIII had a spiritual director named Père Caussin, a Jesuit who was secretly involved with the Spanish faction and who also tried to use his influence with the king to awaken Louis to the desperate condition of his people. Ironically, it was Cardinal Richelieu who had first introduced Caussin to the king and had appointed him his spiritual director. Caussin's politics were decidedly anti-Richelieu, however. He opposed the cardinal's attempts to reach out to the Protestants, and conspired with the Spanish faction to influence the king to reverse Richelieu's policy of siding with the Protestants against the Hapsburgs during the Thirty Years' War.

Caussin reminded Louis how sinful it was for him to be living in luxury while his own mother, Marie de Médicis, was living in relative poverty in exile after the Day of Dupes. He reminded him again about how sinful it was that his people should suffer through another war of religion. Finally, Louis spent three nights in sleepless agony after Caussin told him that he could not be forgiven unless he performed acts of spiritual love that were completely selfless, without any self-interest, without even the desire to go to heaven. How one can have such control over his own psyche and be assured of his own motivations to that degree remains a mystery, even today. Actually, Caussin had overstepped his bounds, for his doctrine was not something that the Jesuits taught, and they would have been angry to learn what he had been doing. Its origin was traced along the twisting path back to the abbé de Saint-Cyran, who, surprisingly enough, never taught it.

On the feast of the Immaculate Conception, December 8, 1638, Caussin preached to the king on his sinful behavior toward his mother and on the terrible deeds committed in his name during the war. Once again, during Communion as the king knelt at the altar rail, Caussin argued with him, telling him how sinful he had been and how he needed to redeem himself through acts of perfect love. Upset, Louis invited Caussin to argue his case before Richelieu during supper. Even though Caussin arrived early, the cardinal had gotten wind of what the king's spiritual director had been doing and ordered him to sit in a waiting room while he discussed these issues with the king. Richelieu was nothing if not rational, and he argued each point that Caussin had made until the king's conscience was salved. After that, the cardinal had Caussin banished from the court. Saint-Cyran's name came up in the aftermath of the Caussin affair, which only contributed to Richelieu's decision to imprison Saint-Cyran.

Richelieu's final reasons for having Saint-Cyran imprisoned, however, were complex and subtle, as was typical of the cardinal. First and foremost, Richelieu did not trust the convent at Port-Royal and suspected them of heretical leanings, so that the abbot's association with the good sisters was suspect immediately. Port-Royal had taken too many bright young servants of the king, like Antoine Lemaître, a brilliant young lawyer who was a blood relative of the Arnaulds, into its penitential bosom. Moreover, at the abbot's encouragement, one of the Arnaulds—Antoine the younger, later nicknamed the Great Arnauld—like some rabbit-and-hat magician, wrote a treatise arguing that the Eucharist was so sacred that ordinary sinners dare not approach it without undergoing a process of purification and penance. This galled Richelieu, because his own conscience was so touchy, mainly because of his ruthless exercise of power, that regular confession and Communion were essential parts of his psychological maintenance. Then, Saint-Cyran's old friend and collaborator Cornelis Jansen wrote a book attacking Richelieu's looming war against the Spanish Hapsburgs. But what did the trick was that Richelieu tried to buy Saint-Cyran off with a bishopric, that of the abbot's hometown of Bayonne, but Saint-Cyran turned him down. The fact that the abbot had been ready to accept the same offer from Richelieu's enemies was a personal insult to the cardinal, and the cardinal did not brook personal insults.

And so off to prison went the abbé de Saint-Cyran, the newly made martyr of the age. Eventually, Richelieu released the abbot from the dungeon, but only after Saint-Cyran had gone blind, and transferred him to the minimum security prison at Vincennes, where he spent the next few years receiving guests and writing theological treatises. But Richelieu was as concerned about Saint-Cyran's political impact as he was about his religious influence.

Saint-Cyran had come from a wealthy, deeply religious family of Basques, and at the proper time he was sent off to study with the Jesuits.

He learned his lessons well, but after meeting Pierre de Bérulle, he came to see the teachings of his old masters as too easy, too worldly. Along the way, he made friends with a fellow graduate of Louvain, Cornelis Jansen, who had fallen in love with the writings of Augustine and wanted to find the true Augustine, freed from centuries of commentators.

Jansenism was the result—a spiritual movement within the Catholic revival that brought together these two strands. The Jansenists derived their spirituality from that of Bérulle, but did so after a muscular study of the church fathers, most especially Augustine, by using a method of positive theology, which was an intense, even scientific examination of the texts. What they came up with was something similar to the ideas of Luther and Calvin—that is, that without the grace of God human beings were capable only of doing evil, and that the grace of God, given to some but not to others, was irresistible. This kind of grace, which they called "efficient grace," could not be denied and would always produce the desired result. However, it was not given to all, but only to those whom God had preordained to become members of the elect. In this, they were radical Augustinians, close to the Calvinists in their reading of the writings of Augustine.

The entire Society of Jesus looked askance at this. They taught that human beings have the power to do good as well as evil, and that Christ had come to save all men and women and not just the select few. For the Jesuits, otherwise known as Molinists, the human will had the power to choose good over evil, and divine grace, which was nearly ubiquitous, gave aid and comfort to those striving to achieve God's will. They recognized the impact of original sin on the lives of ordinary people, but held that Adam's sin did not utterly bestialize people but wounded them, stacking the deck toward sin.

Because of the Reformers, however, Augustinian philosophy had become chic. It was perfect for times of uncertainty. If you are one of the elect, your future is assured. At that point, the spiritual life becomes less about conversion than about watching for signs of your inclusion. Just what those signs were was up to the spiritual leader, which gave such men and women extraordinary power over their charges. Richelieu had the power of life and death, certainly, but over his followers Saint-Cyran had the power of salvation. Augustinianism was therefore the perfect theology for putting together a holy remnant. Just as Lenin found out that a highly motivated cadre of socialists could change the world without the agreement of the majority, so the Augustinians found that they could create a church full of janissaries utterly devoted to their cause.

In this sense, the seventeenth century was not that different from our own. Softer, more sympathetic Christianity, as represented by the Jesuits and the Molinists, produced an open, forgiving church that seemed to be edging toward a kind of relativism—"laxism," as Arnauld and Pascal called it. This moral and spiritual relativism produced a backlash, a set of new spiritual movements within Christianity that insisted on dividing the world between the saved and the damned. This was just as true for Luther and Calvin in relation to the Renaissance church as it was for the Jansenists in the seventeenth-century church. An Augustinian backlash reformulated the church along Augustine's harsher lines, lines that were more in tune with the uncertainties of the entire century. In a sense, these two versions of the faith could be typified today as "liberal" and "conservative."

The Void

Nature abhors a vacuum.

—ARISTOTLE, *Ethics*

If we knew what it was we were doing, it would not be called research, would it?

—ALBERT EINSTEIN

"Because," it is said, "since childhood you have believed that a box was empty because you could not see anything in it, you have believed in the possibility of a vacuum. This is an illusion of your senses, strengthened by habit, that science must correct." And others say: "Because you have been taught in the schools that there is no such thing as a vacuum, your common sense, which understood the notion of a vacuum perfectly well before receiving this false idea, has been corrupted and must be corrected by a return to your original state." Which is doing the deceiving: the senses or the education?

—BLAISE PASCAL, *Pensées*

The story of the vacuum begins with Galileo. One spring day, he was watching the workmen pump water out of one of the cisterns on his property. He noticed that the suction pump that they were using worked well as long as the water was at a certain level, but that as soon as it fell below that level, the pump suddenly stopped working. Galileo called in a plumber to repair the pump, but the man told him that there was nothing wrong with it, for as everyone in the plumbing trade knew, no suction pump could raise water above eighteen *braccia,* or about ten and a half meters. Intrigued, Galileo studied the question for a time and then wrote about it in his *Two New Sciences.* He drew a distinction between a force pump and a suction pump, and noted that a force pump, which pushes the water from the bottom, could move the water much higher, depending upon the amount of force used, than a suction pump, and that a suction pump, which works by attraction, was limited in the height to which it could raise the water out of the cistern. He admitted that he was confused by this and was full of wonder, and then it hit him: in a suction pump, a vacuum was created above eighteen *braccia.*

This was quite a leap, because for nearly twenty-two hundred years Western culture had been following Aristotle and quoting his little maxim that "nature abhors a vacuum," as if that explained it all. Almost everyone in the seventeenth century believed that it was impossible to create a void, a space empty of matter, because nature so radically preferred all space to be filled that it reacted violently to keep a vacuum from occurring. We must remember that Aristotle was an intensely earthbound man and that many of his odder dicta, like this one, came from his observations of nature. Greek philosophers had long held that like things tend to congregate together and that unlike things tend to repel one another. Thus, the idea of a force of gravity never occurred to them because they observed that heavy things tend to fall down and light things tend to go up. They didn't need to pursue the matter further. The things that go down are mostly made of earth, and so they go to the place where earth is, and the same for water, while air and fire tended to rise to their natural places.

There were five elements: the four just mentioned—air, earth, fire, and water—that are natural to the earth; and the quintessence or fifth element, a particularly rarefied and perfect form of matter, which filled the heavens. The quintessence had the unique ability to execute perfect circular motions, like the motions of the heavenly spheres. Thus, there were two sets of laws in Aristotle's physics, one for earth and one for the heavens, the earthbound laws applying up to the sphere of the moon, and the celestial laws applying beyond it.[22]

Nevertheless, Galileo's experiments touched off a flurry of new questions and new experiments. If a vacuum could indeed be created in a glass cylinder, would the glass be completely empty of air or only partially empty? Just how empty was empty? And if it was truly empty, would sound be transmitted through it? Would light? In all of Galileo's experiments, he didn't notice that it got any darker in the cylinder with the creation of a vacuum. On the other hand, if sound could not be transmitted through the vacuum and light could be, then sound and light must be very different things indeed, which proved to be true in the end.

But in the seventeenth century, people were still wondering if a vacuum could be formed in truth, and if so, what kind of force would be needed to produce it. Evangelista Torricelli in Florence decided that the use of water columns was simply too unwieldy and began to use columns of mercury to run his experiments, which brought the whole thing down to scale for once. In 1642, he filled a glass tube with mercury and then stuck his thumb over the bottom of the tube and immersed the bottom end into a bowl of mercury, and then removed his thumb. The mercury fell seventy-six centimeters, rather than the ten and a half meters that a water column fell. From his experiments, he concluded that it was the weight of the atmosphere that caused the pressure to fill the vacuum, that we were all living at the bottom of a sea of air, and that the weight of all that air pressed on us so constantly that we didn't notice it. Could it not be this weight, pushing on the surrounding surface of the water, that allowed a suction pump to work? Could it not be the weight of this ocean of air that raised the water to eighteen *braccia* and no more?

Torricelli discussed all this in a series of letters to a skeptical friend in Rome, Michelangelo Ricci, who defended the traditional view with a series of arguments, which Torricelli responded to. And in his responses, he laid out his entire idea. Parts of the contents of these letters eventually came to Père Mersenne by way of François du Verdus, a friend of Roberval's who was living in Rome, though Verdus left out some of Ricci's best objections and much about Torricelli's belief that it was the weight of the sea of air that created the pressure. Mersenne re-created Torricelli's experiments, and after visiting Torricelli in Florence and watching a demonstration of his experiments before a new cardinal in Rome, Giovanni-Carlo de' Medici, he returned to Paris in 1645 and once again tried to re-create Torricelli's experiment but couldn't find enough high-quality glass tubes. A local engineer named Pierre Petit tried the experiment himself and failed, but then passed word of these experiments on to the Pascals, father and son, on his way through Rouen to Dieppe. And that brought the Pascals into the story.

It was 1646, and nearly everyone accepted Aristotle as the last word in the physical sciences. Étienne Pascal, however, was an exception. He was one of the few who had never accepted the Aristotelian cosmology and always thought that it would be possible to create a space devoid of matter if one followed the proper technique.[23] After Petit's visit, Étienne was excited about the possibility of repeating Torricelli's experiment, and waited for his friend to return from Dieppe so that they could work on it together. Happily, they both lived in a city where there were plenty of first-class glassmakers, far better than those in Paris, and after Petit's return they ordered a glass tube four feet long, with one end sealed, and slightly wider than a little finger. They then purchased fifty pounds of mercury, filled a bowl to three fingers, with two fingers of water on top, and then filled the tube with mercury. Petit stuck his finger over the open end and inserted it, finger and all, into the bowl of mercury until it touched the bottom. He checked to see if any air bubbles had slipped past him and settled at the top of the tube. Satisfied for the moment, he removed his finger, and the level of mercury in the tube dropped over eighteen inches. Both men were amazed, and Petit, wondering if they had made a mistake, checked to see if air had gotten in somehow, but found nothing.

At that point, Blaise walked in and joined the conversation. He was skeptical at first and wondered aloud if air could get through the pores of the glass. Petit told him that if that were so, air would continue to penetrate and the level of mercury would go down as they were watching, but it did not. They then slowly raised the tube and were stunned to see that the empty space above the mercury grew larger as they raised it. The height of the mercury level in relation to the mercury in the bowl remained constant, however, until the bottom of the tube hit the level of the water, when all the mercury rushed out and water rushed in, right to the top. This proved that no air had gotten into the tube through the pores, because if it had, there would have been a bubble of air at the top when the water rushed in.

Nevertheless, while Étienne was nearly convinced, Petit remained cautious. Couldn't air have gotten in somehow? Because Torricelli's complete letters had not been transmitted to Mersenne, they did not know that the Florentine had already speculated that the vacuum was caused by the weight of the ocean of air, a notion that Blaise would discover later on his own. Even with his doubts in place, however, Petit set about demonstrating the experiment among his friends and acquaintances in Paris.

At this point, Blaise was twenty-three, a young man full of the energy of youth, an energy that all too often drained away in the middle of his work and left him sickly. His life alternated between rounds of intense scientific investigation and months of languishing in his bed from one of his many illnesses. By this point in his life, he was a slight fellow, with a biting humor and a rude manner, as if this spoiled son of a controlling father had never quite grown up. He could be tender; he could be kind; he could also be quite funny. But when he was on the trail of an idea, he was often unfair and pigheaded. What he lacked in size he made up for with a loud voice and an imperious manner. He was stubborn, hyperintelligent, with a terrible drive for perfection, a man who desperately wanted to wear the humility of Christ but could never quite pull it off. He spent most of his adult life in controversy, both scientific and religious—a fact that surprised no one. Perhaps much of his personality can be explained by his health. He suffered from migraines nearly every day, and numerous other pains debilitated him. Nevertheless, he would make no excuses for himself and would fight on in spite of his illness. And no one could deny his cleverness. But he would need both his cleverness and his arrogance to defend himself during the coming great debate over the existence of the vacuum.

[1646]

Étienne Breaks His Hip

Whose game was empires, and whose stakes were thrones,
Whose table earth—whose dice were human bones.

—George Noel Gordon, Lord Byron

Brooding on God, I may become a man.
Pain wanders through my bones like a lost fire;
What burns me now? Desire, desire, desire.

—Theodore Roethke

In the winter of 1646, Étienne Pascal slipped on an icy street in Rouen and fell, breaking his hip. According to Gilberte's daughter Marguerite, he had been on his way to perform some charitable duty. At fifty-eight, he was no longer a young man, and given the state of seventeenth-century medicine, a broken hip could be very serious indeed. An injury like that required a specialist. But, luckily for Étienne, there were two professional bonesetters living in Rouen at the time: Monsieur Deslandes and Monsieur de La Bouteillerie. Étienne would not let anyone other than these men attend him, for he was convinced of their competence. It was a good choice, for the old man survived and was able to walk, though even he acknowledged later that he had come close to death.

The bonesetters were also pious gentlemen who between them had set up thirty beds for the care of the poor and indigent in their hospital. They treated the poor without charge and took on the job of teaching others their medical specialty without pay. But as guests of the Pascal household for over three months, they came not just to set bones, but to make converts. Needless to say, their influence over the Pascal family changed everything.

Until that time, the Pascals were pious but not fervent. As an *honnête homme,* a cultured gentleman of scientific and philosophical interests, Étienne lived his life by a code of rationality and honor, and suspected extremism in any form, especially the kind of antirational piety that he saw brewing in the Catholic revival, preferring to live his life as much by Michel de Montaigne as by Jesus Christ. Attending to his religious obligations as a matter of duty, he was more interested in intellectual discussions than he was in prayer. In a later century, he might well have been an agnostic.

Nevertheless, the Catholic revival was everywhere, like Christian Huygens's "lumeniferous aether," and in such an environment the most extreme forms of Catholic piety would often seem like heroism. One day, the pastor of their local church, once a member of Bérulle's Oratory, in a fit of pious poverty renounced his benefice, his rights over the income of the parish, and put on the rough homespun of a hermit. Soon afterward, one of Étienne's friends and colleagues, a fellow bureaucrat working for the king, told the Pascals quietly about his own conversion to reform Catholicism, and how it had changed his life. Everywhere Étienne turned, someone was getting religion.

The idea that his colleague had been converted from one kind of Catholicism to another, as if changing religions, was puzzling. It was a fairly new phenomenon, for the Catholic Church had always prided itself on its unity: didn't the Apostles' Creed call it "one, holy, catholic, and apostolic"? Catholicism had an immense capacity to contain within itself a wide variety of spiritual flavors, because in the past, whenever there was tension in the Body of Christ, the Catholic Church had spawned a new religious order. Protestants had changed that by their tendency to split into ever-smaller denominations. In the seventeenth century, however, this Protestant tendency had leaked into Catholicism, so that movement from one group of the faithful to another, from a more secular variety of Catholicism to a more monastic one, became a secondary conversion, a translation from a debased Catholicism to the true faith—Augustinianism at its best. The Pascals found themselves surrounded by fervent Catholics, and by this new enthusiasm that boiled off the streets of Rouen like summer heat. As part of the unconverted, they were nearly heathen.

Then, in the nearby town of Rouville, an evangelical pastor came to town and preached the new piety at all the Masses. The priest's name was Jean Guillebert, and he came equipped with a shiny new doctorate of theology from the Sorbonne. For years, he had followed the medieval tradition of taking a benefice from the parish without actually doing anything for the people and used the money to support his theological studies. But while at the Sorbonne, he had met the abbé de Saint-Cyran and picked up the new piety from him. From that day on, he became a priest in earnest. He actually moved back to his parish and started caring for the people.

This was the man who would convert the entire Pascal family. Apparently, his sermons were entertaining enough, for he drew crowds from all over Normandy. According to Gilberte, Guillebert was filled with an admirable piety and preached the finest sermons.

Now, however, closing the circuit, there were two young men sitting in the congregation in Rouville, taking in every word—two men who happened to be the bonesetters of Rouen, Monsieur Deslandes and Monsieur de La Bouteillerie. And when they arrived for their extended stay at the Pascal house behind the monastery of Saint-Ouen, they were on fire. In the quiet hours, they held long conversations with the family, and if Étienne was not ready to leap from his bed and run off to Rouville, Blaise certainly was. He read Saint-Cyran's *Réformation de l'homme intérieur* (Reformation of the Interior Man) and was mesmerized. Perhaps he was looking for a spiritual life that was as challenging as his science. In the end, Blaise caught fire along with everyone else.

Who knows what people see in a belief that moves them? Whatever caused it, something clicked on in Blaise's psyche, and from that point

on his heart went to war with his mind. There was much that troubled
Blaise in Saint-Cyran's book—as much as what excited him. In one passage, he read that Jansen believed that scientific curiosity was nothing
more than another kind of sexual indulgence, and this agonized him.
Suddenly, the thing that had given Blaise his identity, his greatest joy in
a life of pain, had become a wickedness. How could he seek the salvation of his soul under these conditions? How much of himself would he
have to give up? Everything, it seemed. A shadow fell on his spirit that
would never lift.

Two strange attractors have emerged repeatedly in the long intellectual history of Catholicism; these two traditions, like gravity wells, have
drawn adherents to themselves, becoming fashionable in their turn, white
hot for a time, then cooling off like aging stars, only to be replaced by the
other tradition. Those who seek to find God in all things, who have a general faith in reason, who believe that the world is a wide and good place
and that people are reasonably decent, end up orbiting around the ideas
of Thomas Aquinas (ideas known collectively as Thomism) and his Christian reclamation of Aristotle. Those who seek to find God outside human
experiences, who distrust reason, who think that the world is a shipwreck
and that people are no damn good, usually orbit around the ideas of
Augustine. The more positive Thomism had no problem with Christians
engaging in scientific study. Aquinas himself wrote his *Summa theologiae*
in order to show that Christianity was not antirational, as opposed to the
more advanced Aristotelian science of Islam. Perhaps it was a sign of the
times, or perhaps it was an indicator of the younger Pascal's quirky personality; in any case, he chose the one Catholic theological tradition most
likely to set the two most important parts of his life into combat.

Blaise was then twenty-four years old. He was beyond the age at which
a man strikes out on his own, but in the Pascal family such an action was
unthinkable. Old Étienne had a plan for everyone, especially his only son.
By that time, Blaise had done everything his father had expected of him;
he had achieved preeminence in the scientific world that few had ever at-
tained. His work on conic sections; his arithmetic machine, the Pascaline;
and his experiments on the vacuum—these had made him famous. And

thanks to Marin Mersenne, Blaise's name was known to the best minds in Europe. But all of this was according to his father's plan. What part of his life was Blaise's own? What part of himself was his? It was tragic that at this precise moment in his life, while seeking the truth of his soul, while brawling through that exquisite moment of vulnerability that young men wrap themselves in like a battle flag, he came upon the ideas of Cornelis Jansen and the abbé de Saint-Cyran.

Blaise continued his pious reading and from there plunged into works of theology, ever deeper into the writings of Jansen, Saint-Cyran, and Antoine Arnauld, who had become their great apologist, and bit by bit his conversion took hold. As Blaise was caught in his spiritual reading, he fretted about his life. He began to wonder if what he had been doing, his best accomplishments, was an act of supreme narcissism and a kind of attachment to the world. He had spent his entire young life spinning out variations on the strictest form of mathematical reasoning, and yet this very reasoning was now suspect. Mathematics was the most precise of the sciences and the one antidote Christian thinkers had in their arsenal to fight the skepticism that was rising like a fume from the *libertins érudits*. The seventeenth century was becoming an age of suspicion, because many of the old forms of reason were dying, and many thoughtful people, instead of trying to build up something new, wanted to tear everything down.

Then there were those, led by Descartes, who were casting about for something radically new to fill the gap. How could they build a new metaphysics that possessed both the certainty of mathematics and the scope of Aristotle? This was the question that Descartes was sure he had solved with his new method, one that cleverly wove in the very skepticism of the new thinkers and yet arrived at a new and more certain metaphysics for a new age. But even these attempts to shore up the dike of reason were being called into question.

Blaise, however, embraced the skepticism by acknowledging that human science, for all its power, was deeply flawed. He rejected Thomism and accepted Augustinian pessimism about human reason, perhaps because he was all too aware of the destructive power of sin. His whole life could be explained by that power. Sin and pain accounted for much of his story: his childhood disease, his mother's death, his father's flight from Paris, the terrible peasant uprisings he saw all around him in Normandy, the poverty, and the disease. Had Adam not sinned, none of these things would have happened. When sin entered the world, it brought death along for the ride and, with death, disease. Here, perhaps, was a more satisfying explanation for the realities of the world than mathematics.

Much later, in his *Pensées,* he placed humankind on an open field between the angels and the animals, noble and wretched at the same time. Wretchedness was the product of sin and was constitutional, a part of human nature, whereas reason, also constitutional, was the product of God's grace. But reason could plumb reality only so far, whereas wretchedness sank clear to the bone. The only way to understand the deep truths of life would be through the reasons of the heart, and not the reasons of the mind. We must make use of the nonrational parts of ourselves if we truly wish to understand. A mathematician he remained, but he never trusted mathematics the way Descartes did. In those three months, Blaise Pascal set himself on a course that would make him the great syncopation to the coming modern age, and for all his personal suffering, his writings lent depth and grace to the coming centuries. Years after his death, his ideas haunted the great minds—Voltaire especially, who despised Pascal's trenchant Catholicism and yet could not deny the brilliance of his mind and the sweetness of his French.

And so, Blaise was converted, at least for the time being. His faith would wax and wane over the years. First, with the help of the bonesetters, he set to work on Jacqueline, whose faith would not wax and wane, and then the two of them set to work on Gilberte. Étienne remained skeptical, but slowly drifted in their direction. What was he to do? His children were his life, and whether he wanted them to or not, they had struck out on their own. All his adult life, he had been leading them; now they were leading him.

Still, Blaise was not entirely converted, not quite ready to give up his science. He had devoted too much of himself to it to do that. That would have to come a few years later, after the death of his father and after the night his heart was seared by divine fire.

The Showman

The show must go on,
The show must go on
Inside my heart is breaking
My make-up may be flaking
But my smile still stays on.

—QUEEN, "THE SHOW MUST GO ON"

Live to be the show and gaze o' the time.

—SHAKESPEARE, *Macbeth*

S cience in the seventeenth century was a serious business, but it was also entertainment. Any serious researcher, if he found something of value, would sooner or later have to make a public demonstration of his experiment, either in front of the king or before a collection of interested nobility, or in a public square where the audiences alternately grumbled and were amazed. A scientist paved his reputation with such stones, because public demonstrations often led to wider controversies and therefore to larger audiences. This had a downside, however. One of the reasons Galileo had such a difficult time with the Inquisition was that he published his dialogues for a general audience and not for the hothouse world of scholars, where his ideas might generate some heat, but not enough to explode. There is always a whiff of sedition when new ideas go public. With the lessons of the Reformation firmly in mind, the hierarchy of the Catholic Church naturally wanted to keep a watch on things, to make sure that nothing radical crept past them.

In January and February of 1647, while his father was yet healing from his broken hip, Pascal spent a great deal of money staging a series of big demonstrations of the vacuum in an open square in front of a glass factory. He had ordered a number of long glass tubes, of various lengths, up to forty-five feet long, and bound the longer ones to a ship's mast. Then, after filling them with different kinds of liquid, he used a contraption made of ropes and pulleys to rotate the mast and inserted the bottoms of the tubes in basins of liquid. In each case, an empty space, apparently a vacuum, opened at the top of the tube as the liquid fell, but not all the way. People applauded, astounded by what they saw.

In a sense, Pascal was reworking old ground by returning to experiments done by Gasparo Berti before Torricelli made his discoveries with mercury, but for the people of Rouen it was all new. As with Galileo, the local Aristotelians buzzed with outrage. Jacques Pierius, from the University of Rouen, dashed off a pamphlet—*Can a Void Exist in Nature?* Air must have gotten into the gap at the top of the liquid, he said, air or some even more rarefied gas, unnamed *spirits* that filled the apparently empty space. Pascal, goaded by their resistance, invited any and all who were interested to come to the glassworks once again for a more dramatic demonstration. On the day, spectators collected in the yard before the mast with two forty-foot glass tubes of liquid tied to it, one full of water and the other full of wine. Naturally, the column of water would fall farther than the column of wine, since water is denser than wine, something that Pascal already knew but the spectators did not. What's more, Pascal had already calculated the difference, and, like any good magician, he prepared his audience to be amazed.

Before rotating the mast with the two glass tubes, he staged a question period with them, asking if wine contained more spirits than water. The audience nodded, looking at one another, uncertain; a few shouted yes, of course it does. Then, Pascal said to them, once the mast is rotated, shouldn't they see a greater space above the wine than above the water? After all, the increased spirits in wine should push it down farther than the water. They agreed to that as well. Then Pascal rotated the mast, but the water fell farther than the wine, in spite of the spirits. The crowd was stunned to silence. A few grumbled that it was a trick. Pascal, not satisfied with their silence, went on: he exchanged the two liquids, pouring out the water from one, the wine from the other, and then reversing them. The result was the same. The Aristotelians grumbled louder, and arguments broke out. *What spirits?* some said. *I don't see any spirits, do you?* The Aristotelians said there must be an even more subtle substance inside the glass. How, they argued, thinking of the power of suction, could such a column of liquid be supported by nothing?

One fellow took the argument to its furthest point by proposing a thought experiment: what if they built a glass flask that was long enough to lie tangent to the earth, a tube that would stretch some 8,856 kilometers, and then stuck the bottom of the flask into a canal? Could that bit of emptiness at the top hold up all that water? At this point, Pascal didn't know, and he puzzled over it. Eventually, the question sent him down the right path. He began to wonder if the columns of water, rather than being sucked up from the top, were pushed up from below—the same thing Torricelli had realized a few years earlier, though his speculations had not yet arrived in France. So Pascal performed more experiments, and still they grumbled. No one likes a show-off.

After the demonstration, Pascal carried on a number of public experiments, each with the same general result, each drawing out the same arguments. But these experiments quickly exhausted him, so that after they were done he took to his bed with a migraine and a terrible weakness in his legs. Blaise's illness had grown worse. He developed a rare form of tuberculosis, and his father sent him back to Paris, where the doctors were better. But someone had to take charge of Blaise's care. Gilberte had married her cousin Florin Perier, and they had moved back to Clermont in June of 1641. So in order to preserve Blaise's health, Étienne sent Jacqueline along to care for him.

Once Blaise arrived in Paris, his friends, especially Marin Mersenne and Gilles Roberval, grew anxious that he publish his findings as soon as possible. They were afraid that someone else would grab the credit if Pascal did not publish, and publish soon. But Blaise was ill throughout much of 1647, and his ability to sit at his desk and churn out copy was limited. Roberval fended off Aristotelian criticism as best he could, while Mersenne told everyone he could that Pascal would publish his work quite soon.

This galvanized Pascal, sick as he was, to climb out of bed and write up his results. His concern was a matter of ego, of course, but also of scientific honesty. Pascal had taken great care to perform his experiments correctly, and what would happen if someone who had not taken such care were to publish a paper full of quick and dirty findings? Pascal would be weighing in with too little, too late—would have to battle uphill against poor science. In his final product, *New Experiments About the Vacuum,* dedicated to his father, Blaise described a number of experiments, made with "quicksilver, water, wine, oil, air, etc.," and explained the significance of each one. His purpose was to prove the existence of the vacuum, but then to "leave it to learned and interested persons to test what happens in such a space."[24]

In the *New Experiments,* Pascal describes how he took a glass syringe with a "very exact" piston inside of it and then plunged the entire affair into water. When he pulled back the piston, it created a vacuum with very little resistance, in spite of the fact that the Aristotelians had taught that this would require an infinite force. He goes on to describe how he later used a bellows to create the same effect, and then how he took a glass tube, forty-six feet long, full of red wine, and stuck it into a tub of water, removed the plug on the bottom, and observed how the wine poured out and mixed with the water, turning it pink. The wine kept pouring out until the liquid left in the tube had reached a certain level, and then stopped a number of feet higher than the level of the wine-and-water mixture in the tub. And that was the mystery: why would the wine stop falling? Was it sucked from the top, or pushed from the bottom? He re-created Galileo's experiment by using a piston to suck mercury up through a glass tube, watching it rise until it reached a certain point, and then observing a space open between the top of the mercury and the bottom of the piston. Galileo had interpreted this behavior as owing to the "breaking point" of the liquid rather than to the pressure of the outside air. Pascal then calculated that the eighteen *braccia* reported by Galileo as the breaking point of water, which Pascal translated as thirty-one feet, would turn out to be two feet, three inches when done with mercury.

At the end of these experiments, he was fairly certain that the space above the wine, the water, the mercury, or the oil was not filled with air that had somehow gotten in, either through a mistake or through pores in the glass; nor had any air bubbled up from the liquid, nor was the space filled with a subtle vapor, a "spirit," or whatnot. What he was not certain of was whether the apparent vacuum was a real one, and this was as much a methodological problem as an ontological one. Pascal was not convinced that his experiments had proved the existence of a real, actual void above the glass, only the existence of something that looked and acted like a void. The question remained: what was this empty space?

Back in Paris, Pascal's illness nearly overwhelmed him. He suffered from constant migraines, night sweats, and such weakness that he could barely talk. He rarely bathed because this set off the headaches, and so his body stank. Still, he was able to attend church with some assistance, if nothing else. On Sunday, September 22, 1647, two men, a Monsieur Habert and a Monsieur de Montigny, came to visit the Pascal home while Blaise was off at church, and they spoke at length with Jacqueline. Both men were friends of René Descartes, and announced that Monsieur Descartes had "expressed a strong desire" to meet with Blaise and that he had "the greatest respect" for Étienne and his son because of their many accomplishments. They further announced that Descartes wished to pay a visit the next morning at nine o'clock, if it was not inconvenient, given Blaise's illness. Not knowing what to do, Jacqueline couldn't find a good reason to turn away the great René Descartes, so she agreed, but asked if they could come a bit later. Her brother suffered most in the early morning, she told them. They agreed to arrive at ten thirty instead.

When he heard of the coming visit, Blaise, fearing that he would not have the energy to defend himself, sent word to his friend Roberval and asked him to attend the meeting. Roberval agreed. The next morning, Descartes arrived with quite a retinue: Monsieur Habert, Monsieur de Montigny, the son of Monsieur de Montigny, a cleric, and a few young boys. All of them filed into the Pascals' parlor and greeted Blaise with massive civility and politeness, though everyone in the room knew that Descartes had not come on a mission of mercy but to debate about the void.

Descartes asked perfunctorily about Blaise's health, and then someone brought up the arithmetic machine, which Roberval demonstrated and which they all marveled at. Everything remained pleasant until, perhaps by prearranged signal, someone politely brought up the vacuum, and then there were flourishes and "Monsieurs" all around and everyone knew that they had stepped onto tricky ground, for they had all grown even more polite: smiles, jokes, laughter from both parties as the sides lined up. Descartes, however, had become quite serious. Roberval and the others explained Pascal's experiments to him and asked him what he thought had entered the tube. "Subtle matter," he said, waving off the question. Blaise did his best to respond, but his exhaustion was getting the better of him, and so Roberval tried to carry the burden of the debate himself and plunged in vigorously. A little too vigorously, as it turned out. His arguments remained civil but were too passionate for Descartes' liking. Descartes turned away from Roberval in a huff and said he would treat with Pascal all day on the subject because at least Pascal was speaking reasonably, but he would not speak with Roberval any more. The man had too many prejudices.

Suddenly, Descartes glanced at his watch and remembered another appointment. It was noon, and he had a dinner date arranged in the Faubourg Saint-Germain. Roberval, too, remembered another appointment, in the same part of the city, and so the debate ended, for the time being. Civility returned, and the men joked with one another, though Jacqueline noticed that the jokes smoldered behind the smiles. Descartes offered Roberval a place in his carriage, but before leaving he announced to Blaise that he was unhappy with the discussion and wished to return the next morning at eight o'clock. Presumably he had forgotten about Blaise's early-morning difficulties, but then he turned and the men set off. After dinner with Monsieur d'Alibray, Roberval returned to the Pascal home to sort out the events of the morning, and admitted that his relationship with Monsieur Descartes had not improved on the road.

Descartes arrived the next morning. Blaise's friend Monsieur d'Alibray expressed a wish to come and to bring their mutual friend Monsieur Le Pailleur, but Le Pailleur didn't show up, because, as Jacqueline put it, "he was too lazy to come over to visit." Ostensibly, Descartes came to give Pascal some advice on his health, but in the end he said little other than to instruct Blaise to follow Descartes' own regimen: sleep late and eat plenty of broth. At that point, Jacqueline excused herself to prepare Blaise's first bath. Baths had given him such a headache in the past that he had stopped the practice, which was not a good idea, and Jacqueline wanted to make sure that his bath was not as hot as it had been. With that, the chief witness to the events left the room.

As voyeurs from four hundred years in the future, we can only imagine the buzz of voices from the drawing room, rising and falling, as the debate continued. Was it a replay of the previous day's conversation? Had they come to some kind of reckoning between the plenists and the vacuists? Likely not, for Descartes was later heard to say that Pascal had "too much vacuum in his head." While Jacqueline was still busy with the bath, Descartes left, apparently as polite as ever, after praising Pascal's arithmetic machine, which heartened Blaise, but Jacqueline thought it was merely a "formula of politeness." That night, Blaise merely had night sweats and insomnia, which was an improvement.

Pascal expected his *New Experiments* to set off a reaction, but this happened quicker than he imagined. He was at odds with Descartes, he knew that, but this time the reaction came not directly but through Descartes' old mentor at the Jesuit college at La Flèche, Père Étienne Noël. By this time, Noël was the rector at the College of Clermont in Paris and was a highly respected philosopher. Like Descartes and Aristotle, he believed that there was no such thing as empty space, because the entire universe was filled with matter. There is only existence and nonexistence, and nothing in between. The world for them was no dead machine but a living organism, a creature of God. If a space appeared to be empty, it was actually filled with material too subtle to be seen. Père Noël relied on the authority of Aristotle to make his point, an authority whose ideas had survived for nearly two thousand years. Pascal, on the other hand, believed that only by experimentation, by controlled seeing, and by reason, the pinnacle of which was geometry, could anyone arrive at scientific truth. He was not above appealing to authority when he needed to, but he wouldn't rely on it to silence debate. Besides, an appeal to authority was a creature of theology and not science. But what was most important to Père Noël was that statements about the world should never conflict with sound theology, and so appealing to authority was legitimate. Pascal was himself a deeply religious man, but he did not want to take on religious questions in his physics.

The skirmish between Pascal and Noël occurred in an exchange of letters between October and December of 1647, an exchange that grew hotter with each salvo. At first, Noël praised Pascal for the ingenuity of his experiments, but then he took issue with the entire notion of a vacuum. After all, he said, light passes through these spaces and is refracted, so there must be something there. No, Pascal responded, the refraction Noël mentions took place at the boundary of the glass tube, and not in the space above the mercury. As to the fact that light passes through empty space, the mystery may well lie in light itself, something no one in the seventeenth century had any hope of understanding. "If our knowledge of it were as great as our ignorance of it, we might perhaps know that it would exist in a vacuum with greater brilliance than in any other *medium*," Blaise wrote to Noël.[25]

The Aristotelians, Noël included, were as busy as ants trying to find the substance that filled the space above the mercury. Some said it was fire, others that it was the ether; some said that it was the same substance as the sky. Pascal thought that this was all in their heads, and for all he knew it was an empty space. He saw no reason to presuppose that an invisible, odorless, tasteless, undetectable *something* had filled the area. Why not call it empty? For there seemed to be no other reasonable alternative. Pascal then argued that empty space was not nothingness; rather, it was something between nothingness and real physical bodies, a midway point, a space devoid of material things.[26] This was truly a new thought—pure extension—the first tentative step into modern physics.

Noël received Pascal's letter on the evening of October 31, 1647. Within a week, he fired off a second letter, this one a little nastier than the first. He turned on Pascal's idea that empty space was somewhere between physical things and true nothingness, saying that such an idea was absurd. He returned once again to his original propositions. Empty space could not act as a medium for light, and besides, according to Aristotle, everything that existed had to be either a true substance or an accidental property of a substance. For example, a red ball is a substance because it physically exists, while the color "red" is an accidental property of the ball. Pascal waved aside the entire distinction, and even brought up a few Jesuits to back him up. He also mentioned Pierre Gassendi, another of Mersenne's group, who argued that empty space was filled with God, and was therefore not truly empty. Before creation, God existed in an infinite, unmovable empty space, and at the moment of creation, he filled it with light. If everything in the universe disappeared, Gassendi argued, we would be left in an emptiness, an infinite space that things exist in and move in, but that itself remains unmoving.

In his second letter, Noël tried to shift the entire argument onto theological grounds by saying that this empty space was really the immensity of God. Pascal, seeing quicksand, backed away. He didn't know what "immensity of God" meant precisely, and after what had happened to Galileo, it was better to avoid the whole question. He took a mystical position rather than a theological one, referring all such theological terms to the great mystery. The mysteries of God are "objects of adoration," he said, not disputation, and therefore they should not be discussed in arguments about physics. Noël was nonplussed. How could Pascal not want to dispute these theological and philosophical concepts? The Jesuit was a philosopher first and last, an Aristotelian and a Cartesian. Debate, argument, disputation—these were his meat, his drink. What was experimental science compared to these things? Moreover, as a teacher he felt that it was his duty to be the gatekeeper, to guard the nation from irresponsible ideas. Like Christoph Clavius in the Galileo case, Père Noël was a natural conservative, for he was the defender of the tradition. Pascal wanted to step outside the normal course of things and to reason in a way that was utterly new.

Pascal did not respond to Noël's second letter, which puzzled many of his friends. Was Pascal abandoning the field to Descartes? In a letter to his friend Jacques Le Pailleur, Pascal explained that he had heard through other sources that Père Noël had been concerned about Pascal's health and had begged him not to bother with a second letter, that the Jesuit had intended the letters to be a private conversation and not a public debate. In other words, he had offered Pascal a truce. Pascal admitted that if the offer had not come from one of the good fathers, he would have been suspicious.

Perhaps Pascal should have been. In January 1648, Père Noël published a book, *The Fullness of the Void,* where he quoted parts of his own letters to Pascal. Understandably, Pascal was furious. Because of the supposed truce, he had kept silent, and many people thought that Noël had won the day, that Pascal had retreated, full of shame. Pascal intended his letter to Le Pailleur to be widely circulated, so he added that he intended to answer the Jesuit's book in a treatise of his own and to conclude "that this space is empty until someone has shown that a matter fills it." The debate moved on to its next chapter.

[1647–1652]

Jacqueline's Vocation

It is true that I am free now to make my own commitments.
It has pleased God, who chastises by granting favors and
who favors us by chastising us, to take away the last
legitimate obstacle that could prevent my making the
commitment I want to make.

—JACQUELINE PASCAL TO BLAISE, MAY 7–9, 1652

While Blaise was rocking the scientific world with his experiments on the vacuum, his sister Jacqueline remained with him in Paris. She was his primary caretaker, seeing to his bath and his diet, sending for the doctor when his health declined, and she was not happy. Jacqueline was arguably her brother's equal in intelligence, and she too had enjoyed her touch of fame. But at this point, her life had become something out of a Victorian novel—an invalid brother, little freedom, no life of her own, few options. All she needed was a dark and creepy moor. The fact that she lived in Paris, a green, promenading city, did not satisfy her; she had ideas of her own about how her life should be. Ever since her brother had become interested in Jansenism, he had carried her along with him, and although his zeal had cooled after they returned to Paris, Jacqueline's had not. While her brother had been out proving the existence of the vacuum, the empty place in Jacqueline's heart had been

SOUTHSIDE CHRISTIAN SCHOOL

STUDENT PASS

DATE: 3/6

STUDENT: Bradley Thomas

Hour: (Circle) 1 2 3 4 5 6 7 8

Assigning Teacher/Room: _____

Destination Teacher/Room: BG

Remarks: _____

ADMISSION
(✓) Excused Tardy
( ) Unexcused Tardy
( ) Excused Absence
( ) Unexcused Absence

DISMISSAL
( ) Appointment Dist.
( ) Sick/Injured
( ) Other

TIME: 9:55

cld-1/25/07

growing, and she longed for the convent, for the life of a holy sister. She kept these feelings to herself for a very long time, but eventually they came to the surface.

Much of what Jacqueline was feeling was true piety. She had been, bit by bit, experiencing the second and, for the Jansenists, the true conversion, whereby the mere churchgoer becomes the fervent daughter of Christ. The rest was a desire to break out and make a life for herself. Gilberte had been a happy wife and mother for years, but Jacqueline never wanted that life. She had already refused several likely young men, but that left her with few options. By default, she was the spinster sister of the great man, the great but sickly man, and the daughter of another great man, and, by convention, she was swamped by family obligations. There were really only two paths available for a single woman from a bourgeois family: she could stay home and take care of her sickly brother, or she could enter a convent and be free. It is hard for contemporary people to see the life of a nun as a symbol of freedom, but that is what it was for many women, from the earliest days of Christianity. The path of the holy virgin, or the consecrated widow, was the only way a strong-willed young woman could free herself from the control of her family. Jacqueline had been growing in her commitment to the Jansenist vision of Catholicism for years, and desired the freedom to pursue it. Just before she and Blaise left Rouen, Jacqueline had received the sacrament of Confirmation, which she prepared herself for by reading a number of tracts by Saint-Cyran, after which she was never the same. Gilbert attributed the change to the Holy Spirit.

After Jacqueline and Blaise returned to Paris, they visited Port-Royal as often as they could. Blaise's health improved through the year because of the treatment he was receiving from the doctors in Paris, and so he was able to publish his treatise on conics and continue his work on the vacuum. In between sick days and work days, they attended Mass at the convent and listened to the sermons of Père Singlin, Saint-Cyran's successor at Port Royal, and Jacqueline's certainty about her vocation grew. One day, she pulled Blaise aside and told him about her desire to become a nun at Port-Royal, and he supported her at once.

So it was puzzling for him that while Jacqueline had been so easily accepted at Port-Royal, his own conversations with Père Singlin and Mère Angélique did not go well. He and Jacqueline spent time talking about spiritual matters with them after Mass, and when they asked, he explained his own religious experience to them, but they reacted with doubt and suspicion. Blaise was a scientist, a mathematician, and they did not trust him. His interest in science was all too worldly, a taint on his piety, an attachment that came between him and his jealous God. They both recognized Jacqueline's devotion as pure, as an unspoiled piety, for she did not try to understand God in her mind, but accepted him in all his divine irrationality. The truth of God was beyond human reason, and Blaise had not freed himself from that terrible habit of thought that wanted to know the world as it stood to the senses, apart from God. But Blaise could not see this. He was cheerfully ignorant of the source of their concern, and would have sunk into a depression if he had known. Never once forgetting what Jansen had written about scientific study, he felt torn between his desire for a rational understanding of the world and his desire for God. He would not abandon reason so blithely, so he strove to use it to bolster his faith. Reason itself, he told them, shows the truth of Saint-Cyran's principles, and the foolishness of his opponents. Père Singlin and Mère Angélique squinted back at him in silence. Here was Blaise the intellectual with all his power of argument arrayed on their side, and yet those very powers were suspicious. The fact that Mère Angélique's own brother Antoine was as intellectual as the young Pascal, as trusting in reason as he was, did not seem to matter. The sisters of Port-Royal took Jacqueline to their hearts at once, but to Blaise they remained cool.

In 1648, the year after she and Blaise returned to Paris, Jacqueline wrote a long letter to her father, still in Rouen, asking for his permission to enter Port-Royal. Blaise ardently supported her desire, at least at this point, even though he would have been the one most affected by her departure. In her letter, Jacqueline said she was willing to abide by her father's decision but that she was also determined to have her own life. "Since ingratitude is the blackest of vices, everything that comes close to it is horrifying," she wrote. Denying her father's will would have been a sin of ingratitude, waving off the fact that he had given her life and raised her into adulthood. Paternalism was the norm among the bourgeoisie in the seventeenth century, and children owed their father more than just respect; they owed him obedience. Much of the first part of the letter was filled with protestations of her profound obedience, how she had always obeyed him and how she believed that it was God's will, "whom we must consider in all matters," not only that she obey her father but that she follow her heart. She had done her duty by Étienne, and now she wanted out. "After all this, Father, I can no longer doubt that you wouldn't do me the honor of agreeing with me and granting me my request."[27]

Jacqueline was asking her father if she could make a retreat to Port-Royal and there test her desires to enter that convent, to see if she indeed had been called by God to that life. She must have already known what her father's reaction would be, for even at this point she puts the decision back into his hands. "On the other hand, if God leads me to understand that I am right for this place, I promise you that I will put all my energy into waiting serenely for the moment you would like to choose for his glory."[28] She must have known that Étienne, the possessive father who sacrificed much for his children and then gathered them around him like chicks in a thunderstorm, who even when he was on the run from Richelieu's police kept control of their lives, would resist any thought of her entering Port-Royal. Marriage was one thing; it brought new children into the family. For her to enter the convent meant a kind of death; it would be a parting that would take her outside his influence. His daughter would pass beyond him into the cloister and be lost to him forever. Toward the end of that year, Étienne finished his term of office in Rouen and returned to Paris, where he gave Jacqueline his answer. Deeply upset, tearful, he told her that at sixty-one he was an old man and that he needed his children about him in his last years. He could not part with his youngest daughter, and would not.

He finally agreed to allow Jacqueline to keep a kind of faux cloister inside the Pascal home, never going out, except to Port-Royal for Mass, and receiving no visitors. Obediently, she took to her room and followed her vocation there until her father's death.

The Great Experiment

When men have realized that time has upset many fighting
faiths, they may come to believe even more than they believe the
very foundations of their own conduct that the ultimate good
desired is better reached by free trade in ideas—that the best test
of truth is the power of the thought to get itself accepted in the
competition of the market, and that truth is the only ground
upon which their wishes safely can be carried out. That at
any rate is the theory of our Constitution. It is an experiment,
as all life is an experiment.

—OLIVER WENDELL HOLMES

Every experiment is like a weapon, which must be used
in its particular way—a spear to thrust, a club to strike.
Experimenting requires that a man know when to thrust
and when to strike, according to need and fashion.

—PARACELSUS (PHILIPPUS AUREOLUS THEOPHRASTUS
    BOMBAST VON HOHENHEIM)

P ascal fretted throughout the controversy. Even in his own
mind, he had not proved the existence of the vacuum. His
experiments, dramatic as they had been, could always be challenged as
mistakes, as uncalculated effects, or just as sloppy workmanship. Since

Pascal had always played the devil's advocate in the debates with his father and Monsieur Petit, he could not get past the possibility that he had created only an apparent vacuum and no real vacuum at all. Père Noël and Descartes were right: the burden of proof was on him. Could there not at least be a limited horror of the vacuum, as Galileo suggested? It is one thing to break with two thousand years of intellectual tradition, but it is another to do so with ease. One other experiment remained to be done, an experiment that, Descartes said in a letter to Père Mersenne, he had already suggested to Monsieur Pascal. Someone would have to perform the Torricelli experiment several times in one day, at different altitudes, which would mean climbing a mountain. If, as Pascal suspected, and as Torricelli had first suggested, we are all living at the bottom of a sea of air, the pressure of that sea would grow less as one climbed higher. A reasonable assumption, but whom to get to do it? Climbing a mountain was impossible for Blaise because of his fragile health. With the weakness in his legs, he could barely walk across town.

Pascal had someone in mind, however: his own cousin and brother-in-law, Florin Perier, who was a lawyer back at the old homestead in Clermont, but who was a physics buff on the side. And he had just the place, too—the Puy-de-Dôme, the mountain that loomed over the old city and was at least several thousand feet high. But would Florin do it? On November 15, 1647, Pascal wrote to Monsieur Perier, detailing the experiment and wheedling him into taking on the task. "I should not interrupt the continual work in which your duties engage you for the purpose of talking with you about meditations on physics, if I did not know that they serve to entertain you in your hours of relaxation and that whereas others might be embarrassed by them you will find them a diversion," he writes with a little bit of humor, a little bit of flattery, a little bit of commiseration. Perier was a lawyer, just like Blaise's father, and Blaise knew how much work went into such a life. Perier must have complained to him at one time about his workload, and Pascal knew that if was going to get the man to take on this experiment, it would be one more burden in his life.

But he also knew that Perier was interested in Pascal's experiments. He had seen the Paris experiments for himself, and he had tried them in

Clermont on his own. If he was not a professional physicist, he was at least a fellow traveler. In his letter, Pascal tells Perier of his suspicions, that Torricelli had been correct, that the supposed horror of a vacuum had been the result of air pressure and not of some metaphysical principle, as if the insensate air could feel passions—horror, aversion, attraction—as if it were human and had a soul. But, he goes on to say, "for the lack of convincing experiments, I dared not then (and I dare not yet) give up the idea of the horror of a vacuum."[29] Pascal, for all his feistiness with Aristotelians and Jesuits, was a conservative man at heart and, unlike Galileo, did not joyfully leap into the new thinking, but plodded along until he was able to convince himself of the truth of the experiments.

Pascal had no doubt that Perier would perform the experiment. The only problem was his busy schedule. He was so certain of Perier's participation that in his letter he related how he had already told many of his Parisian friends, including Père Mersenne, all about the coming experiment, and that Père Mersenne had immediately sent letters to his correspondents around Europe. Within a few weeks, Pascal assumed, practitioners of physics in Italy, Holland, Poland, and Sweden would have heard about what Pascal was up to. More than a little bit of pressure on dear Cousin Florin.

But the experiment may indeed have been an imposition on the already overburdened Perier, because he was not able to get it done for another ten months, all the way into the next September. He traveled a great deal in his work, and there were weather problems, and this, that, and the other—which must have sparked some anxiety in Pascal, since Mersenne himself had written to others encouraging them to try the experiment for themselves, and to tell them of his plan to try it himself, if only he could get some decent glass tubes. Through much of the time, the weather had been bad on the Puy-de-Dôme, first snow, then rain. Finally, on September 19, 1648, Perier carried out the experiment, and wrote this account of it:

The weather was chancy last Saturday, the nineteenth of the month. At around five o'clock that morning, though, it seemed to be clear enough; the Puy-de-Dôme was visible at that time, so I decided to give it a try. Several important people in this city of Clermont had asked me to let them know when I would make the ascent, and so I informed them. Some of these people were clergymen, while others were laymen. All of these men were leaders of the community, not only professionally, but intellectually. I was delighted to have them with me in this great work.

That day at eight o'clock, we met in the garden of the Minim Fathers, which has the lowest elevation in town, and began the experiment in this way: First, I poured sixteen pounds of quicksilver that I had purified during the preceding three days into a vessel, and then took several glass tubes of the same length, each four feet long and hermetically sealed at one end and open at the other, and placed them in the vessel and performed the experiment in the usual way. I found that the quicksilver stood at twenty-six inches and three and a half lines above the quicksilver in the vessel, for all the tubes. I then repeated the experiment two more times while standing at the same spot, and found that the experiments produced the same results each time—same horizontal level, same height.

After that, I attached one of the tubes to the vessel and marked the height of the quicksilver, and left it there. I asked Father Chastin, one of the Minim brothers, and a man as pious as he is reliable, a man who reasons well in these matters, to be so kind as to watch if any changes should occur during the day. Taking the other tube and a portion of the quicksilver, and accompanied by a small crowd of gentlemen, I walked to the top of the Puy-de-Dôme, around 500 fathoms higher than the monastery, where upon doing the experiment, we found that the quicksilver reached a height of only twenty-three inches and two lines, a difference of three inches, one and one half lines. We were ecstatic with wonder and delight, and to fulfill our own curiosity, we decided to repeat the experiment. So I repeated it five times with great care, each at different points on the summit, one time in the shelter of a little chapel standing there, once in the open air, and once more in the rain and fog, which came and went. Each time, I carefully evacuated any air that might be in the tube, and in each case, we found the same height of quicksilver, which satisfied us completely.[30]

They had done it. On the way down the mountain, Perier performed the experiment two more times, and then once they returned to the convent of the Minims, they checked with Père Chastin, who said that the level had not budged, "although the weather had been disturbed, sometimes clear and still, sometimes rainy, sometimes foggy, sometimes windy." They did the experiment with the other tubes one more time, and found the same result they had that morning—twenty-six inches and three lines of mercury. They felt that they had proved it, that the *horror vacui* lessened as they climbed the mountain, which meant that it was merely the result of pressure from the sea of air, which grew less as they climbed the mountain. This was proof even Pascal could accept.

Pascal was delighted, and upon receiving his brother-in-law's letter, he repeated the experiment in Paris, climbing to the top of the bell tower in the church of Saint-Jacques-de-la-Boucherie, about fifty meters from the ground, where the mercury level dropped two lines on the glass. He tried it once more in a private home, climbing some ninety steps, and the mercury dropped by half a line. It was done. It was proved. Everyone could see it.[31] Even Descartes—not that he would have admitted it too publicly.

A Skirmish with the Devil

Heresy is the lifeblood of religions. It is faith that begets heretics. There are no heresies in a dead religion.

—ANDRÉ SUARÈS

If there were an art to overcoming heresy with fire, the executioners would be the most learned men on earth.

—MARTIN LUTHER

S aint-Cyran had been dead for four years. He had died in 1643, the same year that his disciple Antoine Arnauld published *De la fréquente communion*, the little tract that started a theological war. It was also the year that Louis XIII died of tuberculosis and the year after the death of Cardinal Richelieu, Saint-Cyran's great enemy. Blaise Pascal received a copy of Arnauld's pamphlet soon after it was published, and admired it greatly. It seemed to connect beautifully with the Augustinian tradition that Guillebert had been teaching him, in harmony with the music of his faith. For Pascal, the universe was becoming ever more strange, intensely mathematical and yet beyond all logic. The packed world of Descartes had given way to a world where empty spaces could be created without much struggle, where the emptiness of the cosmos seemed to go on infinitely. God, therefore, had to be equally strange, and the human powers of reason, for all their greatness, were powerless to plumb his divine depths.

Suddenly, as if by God's will, his new fervor was put to the test. Pascal was still living in Rouen, in his father's house, and would not return to Paris for four more years. There came to the nearby town of Crosville-sur-Cie a forty-five-year-old cleric, a former Capuchin turned diocesan priest, by the name of Jacques Forton, sieur de Saint-Ange, a minor noble who had studied theology and held a post teaching at the University of Bourges. He taught a "new philosophy," as Gilberte Perier put it, "that attracted all the curious." In fact, Saint-Ange had become quite fashionable in his way, having gathered the notice of Cardinal Richelieu's nephew, who had secured the parish for him.

Two young friends of Pascal's, both fervent admirers of Saint-Cyran and of Antoine Arnauld, alarmed by this man's new ideas, came to the Pascal home in Rouen and asked Blaise to accompany them to speak with the man. It didn't take long, however, once their conversation with Saint-Ange had begun, before they decided that the man was a rampant heretic. He was in fact a blatant Pelagian who, according to Gilberte, believed "that the body of Jesus Christ was not formed out of the blood of the Blessed Virgin Mary." Apparently, Jesus's body was made of some special substance created just for that purpose.[32] Moreover, he taught that reason could demonstrate the existence of the Holy Trinity, and that faith was necessary only for those incapable of rational thought. The implications of this were outrageous to any pious young Jansenist, Augustinian to the core. What need was there for revelation? Where were the mysteries of the faith? Was he saying that people could be saved without divine grace? God in heaven! It could not be borne!

The three young men swore to bring this heretic, this devil, to heel. Following the advice of St. Paul, they first privately admonished him, holding long conversations with him and, of course, keeping meticulous records of the conversations as evidence. But he would not budge, so they referred his case to the coadjutor archbishop, who was an administrator—a church bureaucrat and not a fire-eating theologian—and he promptly delayed taking action. Frustrated, the three sent the case to the archbishop himself, an old man ready for retirement, who was not particularly interested in theological controversies on his doorstep. The three pushed and pushed, however, until they got a judgment against the priest, a judgment that pleased no one. The archbishop refused to name Saint-Ange a heretic, but he also refused to allow him to serve as a priest in Crosville-sur-Cie, which meant that the priest had to move on, which he did, and found another parish fairly quickly.

The skirmish was over, and yet Pascal, seemingly the victor, paid a price. His health declined precipitously, and his energy flagged. He had been in the midst of his debates over the vacuum, fresh from his public experiments, and the battle for orthodoxy pushed him over the edge. But the struggle was typical of Blaise Pascal, and spoke eloquently about his version of the faith. He was a theological pit bull. His Jansenism was a thing to be defended, like the vacuum. It was a set of principles, like science, and those who accepted took responsibility for the defense of those principles. In the war with Saint-Ange, Blaise took the offensive, attacking the priest's position vigorously, as if he were a champion blowing a horn for battle. As Saint-Ange quickly learned, those who disagreed with Blaise had better move aside.

Port-Royal and the Clan Arnauld

The brethren asked the abbot Poemen about a certain brother who fasted for six days out of seven with perfect abstinence, but was extremely choleric. Why should he suffer so? And the old man answered, "He that has taught himself to fast for six days and still cannot control his temper should bring more zeal to less toil."

—THE DESERT FATHERS

L ike the Pascals, the family Arnauld came out of the Auvergne and landed in Paris, where they became lawyers at once. The father, Antoine Arnauld (1560–1619), achieved membership in the Assembly of Paris and later became a counselor of state under the soon to be assassinated Henri IV. His anti-Jesuit feelings appeared early on, in some of his short political writing, especially *Le franc et veritable discourse du roi sur le rétablissement qui lui est démandé des jésuites,* which he wrote in 1602. He also represented the Sorbonne in a lawsuit against the Society of Jesus, and pounded out a career-making speech that had them expelled from France for a short time. Years later, after the Jesuits had returned to Paris and Antoine's children were bloodied by their theological battles with the society, the wits of the city referred to Antoine's lawsuit as the "original sin."[33]

The Jesuits were not just any order of Catholic priests. Unlike Benedictines and Cistercians, they did not retreat from the world but lived in it, determined to inject Christianity into the bloodstream of modern life. They wanted to change the world. In the eyes of many, they were suspect because they were connected to the pope by ties of a special vow. They were *ultramontanists,* who supported the church on the other side of the mountains rather than the church in France, and many nationalists saw them as Papists only and not truly French in their hearts. They were the great opponents of Protestantism and of Augustinianism in almost any form, though they honored St. Augustine, the man and the tradition. Where Thomas Aquinas hedged his bets about predestination, the Jesuits did not. They were against it and any form of Calvinist extremism that it implied.

For his own part, Antoine Arnauld the elder was good at two things—arguing before the bar and procreating. He and his wife, Catherine Marion, had a total of twenty children, half of whom died young while the other half lived longer than their father, who died in 1619, seven years after his youngest son, Antoine, eventually nicknamed "The Great," was born. They had six children who achieved notoriety, while three of these six gathered lasting fame. The first was Jacqueline Marie-Angélique Arnauld, the third child of twenty. Like all the Arnaulds, she crackled with wit and intelligence, but also like them, her greatest gifts were also her greatest flaws. Some people possess beauty—she had that, a little at least—and some possess intelligence—she had that as well—but her defining characteristic was her will. She was the kind of woman who, once set upon a path, pursued it to the end, and would not turn back. This strength of will would bring her honor, and in the end destroy her and everything she built.

As the third child of the vast Arnauld clan, a family of courtiers on the rise, she had few choices in life. She could be married to a likely man, not for love but for family gain, or she could enter the convent. Virginia Woolf had not yet been born, and the independent woman of means was still two centuries away. Jacqueline's grandfather Marion had decided that she was meant for the convent when she was still a child, and when presented with the idea, she agreed, but added the condition that they find a way to make her an abbess. She was eight years old in 1599 when she entered the Benedictine abbey of Saint Antoine in Paris. A year later, she moved to the abbey of Maubuisson, whose abbess was Angélique d'Estrées. Typical of the times, the abbess's sister Gabrielle was a mistress of Henri IV and a beauty at court, a woman of great charm and grace, with a touch of infamy. She had *l'esprit,* as the ladies of the court called it, in spades. It was also typical of the times that no one thought it odd that the royal mistress had a sister who was an abbess.

Actually, the abbess was more infamous than the mistress. Gabrielle had reproached her at least once, saying that she was an embarrassment to the family, the "disgrace of our house." A mistress can play the coquette, apparently, but not an abbess. Either way, life was not very difficult in Angélique's monastery, for religious discipline was nearly nonexistent. The ladies, all from wealthy families, spent their days in idle pursuits, gossiping, eating delicate foods, and even engaging in secret liaisons from time to time. Jacqueline Arnauld took to the life like a fish, and on her confirmation changed her name to Angélique to mimic her heroine, her mentor, her abbess. It must have seemed like a perfect life to an independently minded young girl—no family obligations, no parental nattering, and yet all the comforts she could ask for. Jacqueline, now Angélique, spent her days reading novels and Roman history, walking in the woods, and visiting with friends in the city. She ate sweetmeats and pastries, fine delicate cheeses from the country, and drank the best wine. Her family could come and go as they pleased, and everyone thought that the arrangement was perfect.

In 1602, when Angélique Arnauld was eleven years old, her father worked a miracle of bureaucracy. Getting a papal bull for a child to become the abbess of a convent was nearly impossible, and frowned on by everyone, especially as the church was busy trying to free itself from the corruptions of the Renaissance. Nevertheless, he pulled it off. We can only speculate on how he did it, but deception definitely played a part.

Angélique then became the coadjutrix of Port-Royal, but her life changed little.

Over the years, she grew to hate the convent; she hated religious life, and she was furious with her family for forcing her into it. She rarely prayed, and God had little to do with her life at all. Gradually, by the time she was seventeen, she had sunk into a long, stretching depression that had become so much a part of her routine that she rarely noticed it. She had become the poster child for ennui. On the day of her final vows, her father placed the official document of her profession before her, and she signed it, but did so "bursting with spite." But she was in the convent, committed to the life, and there was nothing she could do about it.

In 1607, just before her vows, Angélique took sick and returned home to be nursed by her mother, where she received care, for which she was grateful, along with a great deal of parental advice about her lifestyle in the convent. She ignored that part. Then, in 1608, soon after her vow ceremony, a Franciscan priest came to the convent and preached a sermon that changed her life. No one can track how a single sermon, or even a single chance comment within a sermon, can tilt the balance of someone's whole world, but it is not uncommon. Angélique must have been prepared for it, prepared even by her doubts, her boredom, and her emptiness. What had once been a meaningless life suddenly sparked, and she realized that she could have meaning in her religious life if she chose to create it. Choosing was something that Mère Angélique knew how to do instinctively.

From that day on, she determined to live the life of a nun properly, to reform herself and her monastery. It must have been alarming to the sisters under her, so used to their comforts and their freedoms, to watch as their abbess suddenly got religion. Angélique began to discipline her own life, to fast, to pray in earnest, to deny herself all the little pleasures that had so filled her days before. Little by little, she reformed her house, with the help and advice of Francis de Sales. At one point in a back-handed bid for independence, she thought she might wish to surrender her abbacy and join Francis's new Visitation order, but the bishop of Geneva had long understood whom he was dealing with, the kind of obstinacy that squatted inside the young girl, and gently demurred, so she remained at Port-Royal.

Resistance mounted quickly, even from the best elements in the convent, because few of them saw any reason to change their way of living. Many of them had been dumped into the convent by their families and expected to live the life that their station in French society afforded them. Besides, who was this teenager to tell them to reform? Everyone knew that she held the abbacy by sleight of hand and that she had lived a life no different from their own. They all hoped that this child would outgrow her burst of adolescent enthusiasm and learn to live reasonably. What they did not realize was that *la petite Madame de Port-Royal*'s will outweighed her by a significant amount, and that she had set herself on the course of reform and would not be moved.

Waving away their complaints, Mère Angélique imposed the Rule of St. Benedict in the strictest way. The sisters then had to take seriously their vow of poverty, which meant that they could no longer hold personal property and that they would begin to eat a more austere vegetarian diet, without the sweets and delicate cheeses they were used to. The sisters were required to pray and to live a life in silent contemplation, something they had not done at Port-Royal for years. With time, the complaints died down and the sisters began to see the value of the new regime. Meaning began to leak back into their lives.

At this point, Angélique was in harmony with the rest of the Catholic revival, and those who knew her applauded her efforts, her family included. But then she carried things too far, and the whole family sat up, alarmed. The reforms finally hit *them*. The clan Arnauld did not mind if their daughter, their sister, put discipline into her house. She had been sent there to do her family honor by living an upright, honorable life—pious, but not too pious. Acceptably pious, of course. But good God, not saintly!

On September 25, 1609, Angélique's family arrived for a visit, led by papa Antoine and *maman* Catherine, along with a small crowd of offspring. It had been their habit until then to come and go as they pleased, for after all, their daughter was the rightful abbess of the convent, a position she owed to their efforts. Ordinarily, they picnicked on the wide lawns, played games, and gossiped with the sisters, but that day they arrived to find the gate locked. They sent word that they should be let in at once. A sister carried word to Mère Angélique that her family was at the gate, and she sent word back that they were not allowed to enter the grounds and that if they wanted to visit her, they would have to come to the parlor—a shadowy, dusty little room beside the gate that had been built long before to accommodate visitors and then abandoned as the life inside the walls loosened.

Papa was furious. He was not accustomed to being treated like a servant, like a *stranger!* He would not have it. In the parlor, he told her so. He demanded entrance, and when he was denied again, he demanded again, until it became clear to him that his daughter's will was more than a match for his own. After that, he entreated her, pleaded with her, raged at her, but Angélique would not bend. Meanwhile, her mother wept epic tears, and her older brother, Robert Arnauld d'Andilly, the eldest child, screamed at her, calling her a "monster of ingratitude and a parricide!" It did them no good at all. Angélique was just as upset as they were. She sweated; her heart pounded; she nearly cried several times; but she would not bend. And so the clan Arnauld had to gather up the shreds of their dignity and storm off, their carriage flouncing on the rutted roads. Thus ended *"la journée du guichet,"* the "day of the wicket gate," spoken of by Jansenists for the next few centuries as if it had been a battle between empires.

The Great Arnauld

Common sense is not really so common.

—ANTOINE ARNAULD, *The Art of Thinking: Port-Royal Logic*

The youngest of the Arnauld clan was named after his father, Antoine, and was born in 1612. His father died in 1619, when he was only seven. And so, like Blaise Pascal, he too had lost a parent in childhood. He was, by almost every account, the most intelligent member of that family, and in the end he turned out to be the most ardent defender of Jansenism. He attended the Sorbonne in the 1630s, where he became friends with Jean Guillebert, the man who ended up as the priest in Rouville and who converted the bonesetters, who in turn converted the Pascals, who in turn supported Antoine Arnauld throughout all the long years of controversy. The young Arnauld was an intellectual meteor who had earned the best grades in everything and was the talk of the faculty of theology. But he was also a great deal of trouble.

The clan Arnauld shared several abiding traits: they all had wills of iron, they all possessed a deadly intelligence, and they were all born lawyers. Antoine Arnauld had all three of these traits in extra-large amounts. He was known at the Sorbonne not only for his intelligence but also for his scrappy personality. It is likely that he was one of those people who take the most radical position on everything just so they can have the greatest triumph on the rhetorical battlefield. Arnauld would argue over anything and everything, and it's not surprising that he annoyed quite a few of the faculty. While still in school, in 1638, he wrote a letter to the abbé de Saint-Cyran, who had become the imprisoned martyr at Vincennes, famously oppressed at the hands of the notorious Richelieu, and asked him to become his spiritual director. Saint-Cyran had already become involved with Port-Royal several years earlier and knew Antoine's sisters quite well. He fully understood how valuable this young man would be. In 1641, young Arnauld received his doctorate of divinity, and there were probably more than a few sighs of relief when he was gone. In 1643, following Saint-Cyran's instruction, Arnauld threw his hat into the Jansenist ring by writing a short tract on the sacraments entitled *De la fréquente communion* (On Frequent Communion). The tract created a stir almost at once.

In essence, Arnauld argued that because the Eucharist was the body and blood of Christ, no ordinary sinner dared receive it. One should undergo a strict regimen of penance in order to purify the soul and to avoid the sin of sacrilege each time before receiving Communion. This flew in the face of Jesuit teaching, which taught that Communion was not a reward for perfect moral behavior but a medicine for the soul and a vital pathway to God. Arnauld saw the thrice-damned Jesuit teaching as a species of laxity, soft on sin and soft on sinners, allowing insults to the divine presence to occur by the polluting presence of sinners who dared approach the altar of God. Within two heartbeats, the friction between these two camps sparked a nuclear fire.

The explosion happened this way: There was a lady at court, a fine and cultured noblewoman who had had her life changed by the preaching and counsel of Saint-Cyran, who had become her spiritual director. She had once lived a life of easy virtue, and the abbot had confronted her with her sins. She tried to do the same thing with her friends and told them not to receive Communion unless they were in a state of near perfect grace, that they could not approach the altar of God without first confessing all of their venial sins as well as their mortal sins, all of their faults and foibles, and converting themselves from living a frivolous life. Moreover, they could not be forgiven their sins unless their penitence was perfect—that is, unless their intentions were pure, done utterly for the love of God rather than out of fear of the fire.

Her friends responded, following their own Jesuit spiritual directors, that the Eucharist had been given to the world as an aid to salvation and should not be denied people unless they had cut themselves off from God through mortal sin. Arnauld exploded. He could not abide this, for such a positive view of human beings, such a comfortable understanding of humanity's relationship with God, did not take into account the monstrosity of sin and the depths of the wound that Adam's transgression had cut into the world. To reject the world-bestriding power of concupiscence was to reject Augustine himself, to fall into the grave error of Pelagianism. No true believer could ever approach the sacred altar of God while still immersed in the ocean of sin that was ordinary human life.

And so the war between Arnauld and the Jesuits was on, and would rage for the rest of his life and beyond.

The connection between Jansenism and the Arnauld family had been there for some time, however, and could be traced back to Port-Royal. In 1635, Mère Angélique invited Saint-Cyran to give a series of sermons during Lent. By 1637, the entire community was under his direction. For what the sisters of Port-Royal and their devoted followers had already learned to practice, Saint-Cyran provided a theological superstructure. Once, he told them, Adam and Eve lived in moral and spiritual perfection, but that time had passed and was gone forever because of their sin, and so we sinners live under the influence of concupiscence, that terrible draw toward wickedness. If we die in our sin, it is because we choose to do so. But even our choice is created by the all-powerful God, and therefore we can claim nothing for ourselves. All we can do is seek the kind of humility that edges on humiliation, and spend a life in penitence for our sins and for the sin of Adam. Though most people are predestined for eternal damnation, there are those who, through the saving power of Jesus Christ, are predestined for salvation. No one can know just who is damned and who is saved, but we can read the signs, for there are "signposts in the predestined soul" that are not there in the damned. These are: a perfect surrender to God's will; the practice of sincere piety; sacramental and personal penance; the acceptance of God's grace in all humility; and, of course, submission to a spiritual director.

And there was the rub, the turning point, the place where Jansenism crossed the line from Christian spirituality to cult. Submission to a spiritual director, namely to himself—and he pulled no punches on this—was no longer merely a wise act, an advisable thing to do, a part of a spiritual program. Suddenly it was an essential dimension of salvation. Not even the Jesuits claimed this. Obedience to the charismatic leader, to the concrete will of the director, became the main signpost of God's saving power in their midst.

There have been other charismatic spiritual directors. Some, like St. Francis of Assisi, St. Benedict of Nursia, and St. Ignatius of Loyola, have changed the world. Others have led their hapless followers into the jungle and the draft of Kool-Aid. Charisma cuts both ways, and that is the problem. Saint-Cyran and Mère Angélique were two strong-willed people whose own unseen will to power had become tangled in their desire for spiritual perfection, who were unable to see that such a demand for submission was little more than hubris. It was inevitable that they would sooner or later come to loggerheads with other powerful people, religious and secular both, who suffered from the same hubris. The two of them were powerful personalities indeed, but could they compete with Cardinal Richelieu, and later with Cardinal Mazarin? Perhaps, in their spiritual certainty, they did not see the storm on the horizon.

That storm hit in 1638, when Richelieu had Saint-Cyran thrown into prison. "The judgments of God are a terrible thing," Angélique later told Jacqueline Pascal. "We don't think enough about them. We don't dread them enough."[34] The judgment of God was falling upon them, or at least the judgment of Richelieu. But the cardinal's imprisonment of Saint-Cyran was only the first salvo by forces that were beginning to coalesce around the Jesuits. Perhaps the Jansenists were doomed from the moment that Francis de Sales and Vincent de Paul rejected them, but the tides were turning against the Augustinians. The modern age would reject them and their negative evaluation of humanity as a failed experiment and move on.

As for Port-Royal, a replacement for the abbot soon filled his spot, a man of different character and learning, Antoine Singlin, a former Parisian linen draper turned priest, a holy man who had once been the disciple of Vincent de Paul and later turned to Saint-Cyran. He would later engage Blaise Pascal in long conversations about the world, about science, and about serving God—conversations that would change the young man's life forever.

The Fronde of the Parlement

*If an injury has to be done to a man it should be so severe
that his vengeance need not be feared.*

—NICCOLÒ MACHIAVELLI

The first duty of a revolutionary is to get away with it.

—ABBIE HOFFMAN

Cardinal Richelieu died on December 4, 1642, after years of declining health. The vultures circled, and the lions gathered to pick the bones. Louis XIII was overjoyed—he was free at last—but put on a good face and observed all the conventions at the cardinal's funeral.[35] He took care of Richelieu's family, reaffirmed his will, and defended his reputation at court, to the point of frowning on all of Richelieu's old enemies. One exception was Jean François Paul de Gondi, the future Cardinal de Retz, a born conspirator, who was busy politicking for the job of co-adjuter bishop of Paris, under his own uncle, the archbishop. Though he and Richelieu had circled each other like wrestlers for years, Retz got the job after gallantly sparing the life of Captain Coutenau, of the king's light horse, during a duel. The man slipped in the mud and dropped his sword, and Retz, who still had his in hand, stood back with a salute to allow Coutenau a chance to redeem his sword from the mud. Instead, the captain bowed low to Gondi and offered his apology, which was graciously accepted. Suddenly Retz was in the king's favor.

But this didn't last long, for Louis XIII died on May 15, 1643, but not before appointing Retz coadjuter bishop on his deathbed. When Louis XIII died, there were no signs or portents, though France was in serious danger of civil war. At the moment of the king's death, his heir, Louis, took his place on the throne and became Louis XIV, the Sun King. The next day, the entire royal household packed their possessions onto carts and, with all of their servants and all of their guards, moved to the Queen's palace at the Louvre. This was a tradition among French royalty; the new king, they believed, should not have to live in the house where his father had died. Were they afraid of ghosts?

The people loved the young king for his youth, and they loved his mother, Anne, mainly for her suffering. No one likes a monarch who has a good time, but one who suffers delicately has everyone's goodwill. As they traveled toward Paris, the people lined the streets and cheered them, calling the young king the *Dieudonné,* the gift of God, and praying down blessings on the Queen Mother. At the gates of the city, the procession stopped to listen to speeches given by local government officials and by prominent merchants, and this went on and on. More than likely, the four-year-old king was bored.

Because of his youth, his mother quickly became the regent and ruled in his stead, an unstable business because both the queen and her first minister were foreigners, and though the people loved them for the moment, that could quickly change. French nationalists like the prince de Condé resented their coming to power and schemed voraciously in the background. The duc d'Orléans pretended to be upset by the regency and was ready to go to war over it, but then the queen made him the lieutenant general of France, and he went away happy. The prince de Condé became the president of the King's Council, and even he stopped grumbling for a time.

On Monday, May 18, the Parlement de Paris assembled to register Anne's regency, which they did quickly, with pomp and flourishes and plenty of references to the will of God. The queen immediately called the exiles home, freed the political prisoners, and even pardoned many criminals. Those who had lost their jobs under Richelieu were soon given new employment, and all requests were granted.[36] Three days later, Anne named Cardinal Mazarin to be her chief minister, and no one was surprised, though everyone at court and out of court knew him for what he was—Richelieu's creature—and hated him.

At first, painfully aware of her dead husband's shortcomings, especially his constant deferral to Richelieu, the queen tried to rule in her own right, and rule by Christian principles. Not being a holy man himself, Cardinal Mazarin couldn't allow that. He had other ideas, and immediately set clandestine schemes running through the palace to undermine the queen and to maneuver her power away from her. Like his predecessor, Richelieu, he wanted no restrictions on his own power, but unlike Richelieu, he had little conscience, and sought his own glory over the welfare of the nation.

By that time, Richelieu's taxes had bled the people white, a fact that many pious people, even the queen's favorite Vincent de Paul, had taken great pains to tell her. However, she was no longer in control of the regency and was further hampered by her vision of royal power. Life at court was spent mostly in the search of pleasure—frivolous conversation, rich banquets, plays and concerts, coquetry. Mazarin, to entertain the queen, brought an Italian acting troupe to Paris to stage a musical comedy, *Orfeo,* which the boy king loved and demanded to see again and again, though it cost four hundred thousand livres just to purchase the set and the machinery for the special effects. This did not include the salaries of the players or the cost of transporting them from Italy. The amount of money the court spent on a daily basis was outrageous, and it is telling that the court was oblivious to the effects that their pleasures were having on the populace. While the courtiers tittered over the latest intrigue, the most recent scandal, the people languished in poverty, and the French Revolution inched closer.

In the end, all of her good intentions fell apart in May 1648. The war that was impoverishing the nation seemed to go on and on, while the court spent more money every year on frivolities. D'Emery, Mazarin's superintendent of finance, widely known as one of the most corrupt men of the time, was foraging for new tax schemes. He issued edicts announcing new taxes as fast as he could name them

Up until this point, there had been no popular uprising in Paris as there had been in the provinces. The streets had been clear, the people quiet.

The only thing that had happened was that the tax officers had refused to do their jobs under protest of the new strictures to tax more and receive less, and some members of the Parlement had made speeches against the regent's policies. But Anne would not let this go; she insisted on seeing these protests as the start of an insurrection. And by doing so, she incited the very rebellion she feared. What neither the queen, nor Mazarin, nor anyone else in court realized was that the rebellion had been brewing since Richelieu's day, and that the violent tax revolts that had become commonplace in the provinces were about to visit the capital.

Meanwhile, Cardinal de Retz, who was out of favor at court because of his opposition to Mazarin, used his position as coadjutor of Paris to win the love of the people. From February 25 to March 26, 1648, he distributed thirty-six thousand crowns among the poor. Seeing what was going on, he informed the queen and Cardinal Mazarin of the people's disaffection, and then quietly told the queen of Mazarin's cunning, which earned him no love from the cardinal and no gratitude from the queen.

All of a sudden, news came to the court that the young prince de Condé, the son of the old grumpy lion of Louis XIII's day, had achieved a great military victory in the town of Lens. The court was jubilant. The queen ordered a Te Deum to be sung at Notre Dame. Then she called her council together, and they decided that the celebration would be a perfect opportunity to crush the rebellion in the Parlement. The people would be too busy celebrating to notice that their leaders had been quietly arrested. Just before leaving for the cathedral, she called Comminges, a lieutenant of the royal guard, to her side, quietly informed him of her plans, and placed him in charge of the detachment making the arrests. This decision was the beginning of the Fronde of the Parlement, a rebellion that ultimately failed and merely strengthened the power of the crown, for France was not quite ripe for revolution.[37]

Comminges packed his carriage with four of his guards and one other officer, and together they drove to the street of Monsieur Broussel. Broussel, an old army officer who suffered from gout, was the most vociferous opponent of the king's taxes, and the people claimed him as their great protector. Comminges ordered the carriage to the end of the street, with orders to come at once as soon as he had entered the counselor's house. Then he marched up on foot and knocked on the door. A young boy answered and opened the door for him, and once it was open Comminges leaped inside, holding the door until the carriage arrived. Then, leaving two men at the door, he took the other two guards up to Broussel's apartment and barged into the room, where the man was finishing his dinner with his family. Comminges announced that he had an order from the king to arrest him and take him to prison. Broussel, who had shown great courage in the Parlement in denouncing Mazarin and his policies, was over sixty years old, and trembled at the sight of the lieutenant and his men. He told them that he had taken medicine that morning and that he was in no state to travel, but they would not listen. They grabbed the old man, and at that one of the servants, Broussel's old nurse, ran down the street screaming to the people for help, saying that their protector was being carried off to prison. Suddenly the street was filled with an angry mob. When they heard that Broussel had been arrested, they snatched at the reins, scrambled for the carriage, pulled at the guards. Inside the house, Comminges looked out the window and saw that a riot was beginning. He told Broussel that if he tried to delay any longer he would kill him. Seizing him, he dragged him from the apartment and down the stairs, and threw him into the carriage, while the guards pushed back the people.

With that, the crowd grew more violent and angrier than before; the people pushed and shoved and cursed and threw rocks. The mob teemed all around them even as Comminges and his men tried to escape with their prisoner. He and his men, especially a young page, fought back, but the crowd managed to lay hold of the carriage and overturn it. They would have beaten the men to death had there not been soldiers from the guards standing nearby. Leaping out of the carriage, Comminges pulled his prisoner out the door and fought his way through the crowd. "To arms! Comrades! Help us!" he shouted to the soldiers, and they fought with the people in the streets until the guards brought up another carriage and Comminges escaped with his prisoner. Suddenly, all of Paris boiled with sedition. When the queen heard about the riot, she sent troops through the streets to pacify the people. It didn't work.

The queen's counsel, meanwhile, met in the Palais-Royal to discuss the problem, trying to ignore the sounds of rough singing from the streets below. They laughed, they talked about frivolous things, but not one of them was willing to show what they really felt, what the queen felt—that they were all deeply afraid and had no idea how to solve the problem. Anne, meanwhile, ridiculed the people's anger and told the court that she was not afraid of the people, that she was certain they would do her no harm.

Later that evening, Retz appeared before the queen as an intermediary for the people. On their behalf, he requested once again that they release Broussel and warned her that if she did not do so, the people would recover him by force. But the queen would not bend and ridiculed the idea, sending the coadjutor of Paris back to the people empty-handed. At that point, many in the court could see that the queen's inflexibility would likely get them all killed.

It was in the middle of all this that the Pascal family decided that a visit to the country, to Clermont, was a good idea. They were not attending the queen at that time, because it was Étienne's own class of men who had rebelled. Étienne had returned to Paris after finishing his term of office in Rouen, and when the rioting broke out, his position as a tax judge made him vulnerable to the whims of the mob, though his fellow tax judges were the darlings of the crowd for the moment. But, while the rebellion had been started by men like him, the people had become a force of their own, and their fury could easily turn on the Pascal family as representatives of the system that had held them down for so long. As the riots worsened in the city, Étienne packed his bags once again and, with Blaise and Jacqueline in tow, fled the city for Clermont, their ancestral home. Life at court was not worth this.

As the night deepened, the crowds gradually dispersed and the queen took heart, reassuring herself that there was nothing to fear. She was wrong.

The next day wasn't much better, nor the day after that. What had started out as a strike by government employees ended as a popular uprising. What no one had the foresight to see, neither the members of the

Parlement nor the members of the royal court, was that this was only the first shot, the prologue to a general revolution a hundred years later that would topple the monarchy and set Europe on fire.

The Paris that the members of the royal court imagined as a city of beauty and pleasure had revealed its truest heart. Among the lower classes, it was a city that yearned for revenge; a blood feud was building between the rich, entombed in their privileges, and the poor, desperate and hungry, and there was no solution for it, no solution other than blood. Meanwhile, the queen kept court as best she could, and in the city, the Parlement met and deliberated about the queen. By the end of the next day, they sent representatives to the queen, who met them without pomp in the little gallery. The chief president promised not to deliberate on taxes until after St. Martin's Day, but that was the best they could offer. The queen was not happy, for it was not the solution she desired; she could see that this was only a reprieve. Nevertheless, she recognized that the Parlement was seeking a compromise, and so she ordered the release of Broussel. Vengeance would come in its own time.

Once outside the Palais-Royal, the representatives of the Parlement approached the people gathered on the street and told them that they had secured the release of Broussel. Not everyone believed them. Some announced that if this was a deception, the people would storm the Palais-Royal and pillage it, and then throw the foreigner out!

When the rules of civilization crumble, even for a short time, what is left is the mob. The populace takes on a new personality, darker than its everyday personality, driven more by paranoia than by reason, or even by self-interest. Paris did not settle down after Broussel's return, because the people did not trust the queen, even when she returned their great protector to them. Though their fear was well founded, as it turned out, the driving force of the rebellion was no longer the freedom of Broussel but their mistrust and hatred of the queen. The burghers refused to tear down the barricades and to lay down their arms. The people refused to return to their homes until the Parlement sounded the all-clear. They were all too afraid that the queen meant what she said—that she would avenge herself and her son upon them and their city for their disobedience. Finally, late on the morning of August 28, 1648, with Broussel attending, the assembly published a decree enjoining the people to obey the will of the king and to return to their homes. The decree was passed later that day.

The people listened, the barricades came down, and the crowds melted away. Everyone breathed easily, and the city seemed to return to itself and become a place of charm and grace once again. But this didn't last very long. The guards at the Palais-Royal needed resupply, and, unhappily, their two caissons of gunpowder arrived in the city at just that moment. The people watched as the military procession rolled through the streets of the city, and they became certain that the queen was preparing to strike. All of a sudden, the streets were filled with people calling, "To arms!" And things were just as bad as they had been the day before. The magistrates moved among the people to reassure them, but no one listened. Within half an hour, the city was alive with rebellion once again.

Two months later, the queen, the young king, and the cardinal, leaving much of the court to fend for themselves, sneaked out of the city and rallied troops for their attack on the city. Fortunately for them, the Peace of Westphalia had ended the Thirty Years' War, and the French Army, under the command of the young prince de Condé, was now free to put down the rebellion. His troops surrounded the capital, and all the people's fears seemed to come true. The queen would have her vengeance after all. But the prince was not eager to lead a general slaughter, and never led his troops inside the capital. Meanwhile, the nobles among the rebels quietly negotiated with the Spanish, the queen's own family, to intercede on their behalf. But the people were too French for that, and wouldn't be saved by the hated Spaniards, and so they were forced to submit. Negotiations started; messages zoomed back and forth across the barricades; the people submitted; the queen relented; and suddenly in 1649, with the Peace of Rueil, there was peace at last. The only problem left was the prince de Condé, who had his own army to play with and now felt that the queen owed him. She didn't agree, and the Fronde of the Princes was on.

Adrift in the World

Now a' is done that men can do,
And a' is done in vain.

—ROBERT BURNS

Why, all delights are vain; but that most vain
Which, with pain purchas'd, doth inherit pain.

—SHAKESPEARE, *Love's Labours Lost*

From all blindness of heart, from pride, vainglory, and hypocrisy;
from envy, hatred, and malice, and all uncharitableness,
Good Lord, deliver us.

—*The Book of Common Prayer*

And then I said: O Lord, how long?

—ISAIAH

The Pascal family had fled Paris for the Auvergne in May 1649, and lived there for the next eighteen months at the home of Gilberte and Florin Perier. Undoubtedly, they gleaned whatever news they could of the events in Paris and waited for the fire to die down. Blaise continued with his scientific investigations, and had momentary episodes of religious fervor, but these, too, died down. Gilberte recorded how he made retreats in the country, in spite of his constant infirmity, to help sort out his faith. At home, he took on many of the tasks once reserved for the servants, and refused their help on household chores. He made his own bed, carried his dishes in from the table, even cooked for himself from time to time. Whether that was a bad idea or not, Gilberte never said. He read the Bible voraciously, and said that the Bible was not a text for "the genius," the rational mind, but for the heart. His life was inching toward a fork in the road. His problem was complex—how to reconcile his scientific pursuits with his spirituality, especially when his spiritual directors thought that science itself was a sin equal to lust. He felt, increasingly, that he had to make a choice. The fact that this very idea would have appeared excessive, even ridiculous, to the vast majority of Catholics, especially those trained by the Jesuits, never quite sank in.

Jacqueline, on the other hand, never wavered. She kept her promises to Mère Angélique and remained in her room as best she could, living a life of monastic silence. According to Gilberte, Jacqueline never left her quarters except to go to church or to visit the sick and the poor, all of which were part of the life of the sisterhood. She maintained her contact with Port-Royal through a stream of letters between herself and Mère Agnès Arnauld, the novice mistress and Mère Angélique's sister and closest confidant. Finally, as the Fronde of the Parlement died down, Étienne decided that it was time for the Pascals to return to the capital. They arrived in November 1650 and took residence in a house on the rue de Touraine. The Fronde of the Princes was not quite over with, however, and so the family watched through their second-story windows as troops maneuvered on the streets below and soldiers fired muskets on one another and died.

While all this was going on, Blaise found himself in a war of his own. In the summer of 1651, an anonymous Jesuit attached to the college in Montferrand wrote a treatise attacking Blaise's work on the vacuum. The Jesuits, awash in Thomistic Aristotelianism, naturally abhorred the vacuum, and fought the idea whenever they could. After all, Descartes was their boy and Pascal wasn't. Thomas Aquinas was a saint, and

Pascal wasn't that, either. In his treatise, the anonymous Jesuit wrote that "certain persons, lovers of novelty," had falsely claimed credit for the discovery of the vacuum from experiments performed in Normandy and the Auvergne when in fact these same experiments had already been carried out in Italy and in Germany. Adding acid to the wound, the Jesuit dedicated his treatise to Étienne's successor on the Cour des Aides, the tax court of the Auvergne.

Blaise was furious. There was no doubt about the identity of the "certain persons." After all, the anonymous Jesuit had specified experiments in the very places where Blaise had so famously triumphed over the plenists. Moreover, because he remained anonymous, his attack was particularly cowardly. If you are going to attack someone's character, you should at least do so to his face. Blaise wrote a public response, protesting the attack on his character and on the character of his brother-in-law, Florin Perier. He then outlined the history of these experiments, how he was in a tradition of researchers who doubted the Aristotelian physics, from Galileo to numerous others, and how each one had made his contribution, himself included. The experiments he had performed were his own, and their honesty could be attested to by both his brother-in-law and his father, Étienne—good men of high position.

What was at stake was an entire theory of knowledge. The scientific method was just then being formulated, and Pascal was a part of this. Theological and philosophical knowledge existed as part of a tradition, and so the opinions of great authorities in the past meant something. This was not true of science, for science was a new thing in the seventeenth century, and was still in its infancy. Pascal and others sought to distinguish scientific knowledge from both theological and philosophical knowledge. Cultural questions—who was the first king of France, where the geometers drew the first meridian—are questions of authority and can be solved by reference to books. This is a matter of scholarship. Questions that can be solved only by experiment and by reason—questions in mathematics and physics—cannot be solved by references to authority, and can be solved only by rigorous thought and observation. Père Noël and the other Aristotelians wanted to solve questions of physics as if they were questions of theology. This was not the method taught to Pascal in Mersenne's seminar; it was not the method taught to him by his father.

In the midst of this controversy, Étienne died. His health had been declining for some time, perhaps even as far back as Rouen and his fall on the ice. His long hours of tedious calculation couldn't have helped matters, for he was frequently exhausted in Normandy, and the time he spent there aged him beyond his years. Still, at sixty-four he was an old man by seventeenth-century standards, and his death was not unexpected. He died on September 24, 1651, just as Gilberte was going into labor in Clermont. She gave birth three days later, to a healthy son, and then received a letter from Jacqueline informing her and Florin of her father's death. A month later, she received a letter from Blaise completing Jacqueline's letter. The letter he wrote was quite Catholic. He mentioned his own grief in passing, but the rest of the letter reads like "My Sunday Sermon." For a Christian, all of life should be a dying, an act of sacrifice: "Let us not be afflicted like the heathen who have no hope. We did not lose my father at the moment of his death. We had lost him, so to speak, as soon as he entered the Church through baptism."[38]

At this point, Pascal sets out his idea of the "Two Loves," an idea that would resurface later, in the *Pensées*. God has created humanity with two loves—the love of God and the love of self. The love of God is infinite and all-consuming, while the love of self is finite and has the purpose of leading us back to God. But when sin entered the world through Adam, humanity lost the love of God. Only the instrumental love of the self remained, and "this self-love has spread and overflowed into the vacuum which the love of God has left." In order to recover the love of God, each human soul must die to itself. It must give up the love of self in order to make room for the love of God. And so, dying should hold no fear for the Christian, who has already died in the soul; once that is accomplished, the death of the body is a dawdle.[39]

But in spite of all these brave Augustinian words, for Blaise this was a terrible time, perhaps the worst in his life. Underneath his theology he was in deep emotional pain, and the mention he makes of his grief re-

veals just the tip of his sadness. His father had been the gravitational center of his life from the day his mother died. He had no wife, no children,
few friends; his life had turned around his father's plans for him, and he
had become a great man, greater even than his father. But in all that time,
partly because of Étienne's domination and partly because of his persis-
tent illness, he had never created a life for himself. He was a great mind in
search of a heart, and now, what was there to love? He knew in his faith
that God was the ultimate object of love, but just how does one attain
that goal? The only way to do this, as the Jansenist spiritual directors at
Port-Royale insisted, was to rid himself of all distraction, of every other
love, so that, like the poor of the earth, his soul would have no other place
to go but to God.

But his attachments, it seemed, had their own way of disappearing.
His intellectual achievements had produced nothing but controversy—
bile from the Jesuits, bile from Descartes, bile from the intellectual gate-
keepers of the university. He had followed his father's path, but now his
father was dead, and both of his sisters were setting out to find their own
lives, leaving him alone. What had started in Rouen with Gilberte's departure for Clermont to marry Florin Perier continued on in the death
of his father and his sister Jacqueline's insistence on entering Port-Royal.
One by one, the members of the family he depended on were leaving
him. He was alone in the world.

As soon as Étienne had been laid to rest and an appropriate mourning
time had passed, Jacqueline announced her intention to enter Port-Royal.
After all, she had stayed home out of duty, because her father had asked
her to, because he could not bear to part with her, and because to disobey
him would have been an act of impiety. But now he was dead and her
duty fulfilled. Blaise needed her, too, but he was not her father and did
not have the same authority over her, even if he was the only male left in
the family. Jacqueline was convinced that gender alone should not give
one person authority over another, and that she should follow the calling
of her heart over the needs of her brother. He had servants to help him,
and money enough to live comfortably. Did he need his sister any longer?
Suddenly, there was war in the Pascal household.

Blaise pleaded with Jacqueline not to leave, but she was adamant. He commanded her to stay, but that didn't work, either. At the heart of this growing difference, beside Blaise's fear of abandonment, was a disagreement over Port-Royal itself. Like Francis de Sales and Vincent de Paul, Blaise was shocked by the severity of some of Mère Angélique's directions. While in Paris, Jacqueline had taken an Oratorian priest as her confessor. Since this was the order founded by Pierre de Bérulle, a longtime mentor for the abbé de Saint-Cyran, he was a good choice. The priest had a noteworthy reputation, and no one said anything bad about his advice. He praised Jacqueline's poetic ability, especially after she translated a Latin hymn from the breviary, *Jesu, nostra redemptio,* into French. He told her that her translation was beautiful and that she should seek to incorporate her gifts into her spiritual life. Jacqueline was happy at first, but then suffered a bout of scruples. She fired off a letter to Mère Angélique, who wrote back: "This talent for poetry is not something for which God will ask you for an accounting. You must bury it."[40]

Blaise was aghast at this. The demands of Port-Royal were stricter than even the advice of an Oratorian, one of the strictest orders in the French church. Mère Angélique's advice sounded too much like what Jansen had said about the study of science—that it was another form of lust. Once you begin with the assumption, as Augustine did, that humanity is a bit of "spoiled meat," and that the only righteous thing a human can do is live a life of penance, you will eventually find nothing beautiful in humanity. Everything human that seems good is only an illusion and must be burned out. Everything human that seems righteous is a fancy and must be killed.

In October 1651, a month after their father had died, Blaise and Jacqueline came to an accommodation. Jacqueline signed her inheritance over to Blaise; in return, Blaise promised her a regular yearly income from the estate. These negotiations were between the two of them, for Gilberte had already been given a sizable dowry on her wedding day. Everyone seemed to understand that if Jacqueline entered Port-Royal, she would have to leave her inheritance behind. Nevertheless, Jacqueline was bent on entering the convent and would not change her mind. When Gilberte came to visit at the end of November, Jacqueline pulled her aside and told her quietly that she intended to become a postulant at Port-Royal by the first of the year. She was concerned about how Blaise would react, and so she had told him that she was going for a retreat, when in fact she had no intention of returning. The new year came, and on January 3, 1652, Gilberte, and not Jacqueline, broke the news to their brother. According to Gilberte, "He retired very sadly to his rooms without seeing Jacqueline, who was waiting in the little parlor where she was accustomed to say her prayers."[41]

The next morning, Jacqueline stood in the corridor of the Pascal home, waiting for the carriage to be brought around to take her to the convent. Gilberte saw her and turned away, unable to say good-bye for fear of weeping.

The Feud

You must be able to judge that I am strong enough to go ahead despite you but not strong enough perhaps to withstand the anguish that your opposition would cause me.

—JACQUELINE PASCAL TO BLAISE
  ON THE OCCASION OF HER VOWS

God sets the solitary in families.

—Ps. 68.6

Families, I hate you! Shut-in homes, closed doors, jealous possessions of happiness.

—ANDRÉ GIDE, *Les nourritures terrestres*

By this time, Port-Royal had begun to smell like a cult. At the time, Parisian Catholics would have used another term—heretics—but the social implications were often the same: an increasingly isolated group surrounding a charismatic personality, espousing an exaggerated spirituality. Were they true heretics, or merely extremists? Many of the leaders of the Catholic revival had already rejected them, including some who had once been friendly. And there were factions in France, most notably the Jesuits, who were preparing charges against them.

Their move to Paris had been as lucrative for the community as Mère Angélique had hoped; rich patrons flocked to them, though they had lost the support of the crown. Richelieu had despised them. Although Louis XIII had admired them for a time, the queen had little use for them in the end, and Mazarin had even less. When Louis XIV entered manhood and sat on the throne in earnest, he set about to destroy them. Dark clouds slowly gathered over the convent. Some modern commentators say that Port-Royal was persecuted because they were powerful women, but this is unlikely, since there were many powerful women in France, and some of them were the ones doing the persecuting. Moreover, Port-Royal was merely the core of a wider movement, one that included both men and women, one that advocated both a fearsomely penitential life before a terrifying God and the breakdown of the old ecclesiastical hierarchies in favor of a new egalitarianism. While Mère Angélique ruled her own universe with steel, she increasingly ignored the authority of those over her and promoted equality among the sisters without regard to their station in life on the outside. And so, Port-Royal carried a whiff of republicanism about it, the same kind of republicanism promoted by the Calvinists all over Europe and the New World, the kind that got King Charles of England beheaded.

France was becoming polarized between the Port-Royalistas and everyone else. This may have been the real reason why the crown opposed them: because they were fast becoming a new center of power. But the reason the church opposed them was more theological. Largely because of the Jesuit influence, the tide was turning against Augustinianism, an undertow that would eventually lead to the condemnation of Augustine's teachings on free will and predestination. A rigor that might have made sense twenty years earlier suddenly felt excessive, and the Jansenists had taken Augustinian pessimism one step beyond what French sensibilities would accept. They had gone too far, and had shocked even the most pious souls by their outsize rigor.

One of the marks of a modern cult is that they quickly try to separate people from their money. The argument is that the cause is so important that it transcends all other considerations, and if you truly believed, you would give all you had. This is the same logic that has been followed by religious orders from the beginning of Christendom, and so it has always been difficult to discern a cult from a legitimate religious movement.

For women religious, it has been a common practice to present the house with a "dowry," the same dowry that they would have given to a prospective husband, since on the day of their vows, they would become "brides of Christ." Not to do so would place an undue burden on the house, since religious are not allowed to work a trade to make money. Jacqueline Pascal had entered Port-Royal with pious abandon, and she wanted to enter her new life on an equal footing with all the other sisters. Therefore, she expected to make the same kind of donation that the others had made. It seemed only logical to her that she should give her share of the family fortune to Port-Royal. But suddenly, surprisingly, she bashed into a wall of opposition, not only from Blaise but from Gilberte as well.

Blaise's resistance came from the fact that his health problems had placed him in a precarious position. His doctors and their cures cost money, and with Jacqueline gone, he would have to find some other way to find care. If he handed over a third of Étienne's legacy to Port-Royal, he could be in serious financial trouble. Jacqueline would take a vow of poverty, but Blaise would live it. Of course, he was still angry with Jacqueline for running out on him. In his letter to Gilberte after his father's death, he hinted at his great need. Gilberte had left him for her own life in Clermont; Jacqueline had left him for a life in the convent. Both of them had strong support systems, while Blaise, with all his illnesses, had none. Gilberte seemed to be oblivious to this need, while Jacqueline seemed to resent it. The very fact that Blaise needed her still was a sore point for her. His resistance to handing over her dowry only aggravated that resentment. In all of this, Mère Angélique advised her never to trust people in the world, and as her brother, Blaise, was a man of the world, he was someone who had to be rejected:

> Haven't you learned long ago that you can never trust the affection of creatures, and that the world loves only those who are the world's? Aren't

you happy that God teaches this to you so clearly by the behavior of those from whom you would have least expected it? This should answer any doubts that you might have had, before you leave the world forever. Now you can do so with a bolder heart, because your action is the more necessary. Your resolution must be unwavering, because you can now say, in a manner of speaking, that you no longer have anyone to leave behind."[42]

But Blaise's objections were not the only ones on the table. Florin and Gilberte Perier objected to the fact that these donations might actually deprive them of their legitimate inheritance, because the family fortune was almost exclusively in credits, and it was one thing to hold someone's debt and another to collect it.

None of this was particularly troubling for the convent, for they had the capital to wait on these debts, whereas the Pascals did not. Port-Royal was far from financially strapped. By the time Jacqueline entered the house, they were rich and powerful in a way that few houses of religious women could ever be. They had become a national force, a force that had earned the notice, and the malice, of the powerful. In a sense, Angélique had traded a soft life for power, all too easily done and almost completely done without her awareness. But such a choice was part of her character, she who even as a child insisted that if she was going to be handed over to a convent, she wanted to be an abbess. The love of power is its own form of lust, but if gathered under the auspices of legitimate authority, it can seem like an act of profound spirituality. This was a choice that had plagued many imperfect reformers in the past, male and female alike.

It should not have been too surprising, however, that Blaise would object to his sister's request. Jacqueline had been fighting with Blaise over her vocation for nearly a year, even before the death of their father. Port-Royal had removed her from the circle of the family, and because of its theology of extreme renunciation, it had encouraged her to abandon the world utterly, and if that included the family who loved her, so be it. What was behind this was the Doctrine of Two Loves—that each of us has two loves, the love of God and the love of our own lives, and these two loves are at war. To love God properly, we must abandon everything else, especially all other loves. Even before her father died, Jacqueline had written such ideas in a reflection on the Cross in which she said: "This teaches me to die not only to myself, but also to all the interests of flesh and blood and human affection. That is, I must forget everything about my [family and] friends that doesn't concern their salvation."[43]

Thus, to enter Port-Royal, Jacqueline had to put her old life aside, and that included her brother and sister. From that day on, her only concern for them would be for their salvation. For a pious Jansenist, that meant that she should always work for their own total renunciation of the world. The fact that Blaise had resisted handing over a third of the family fortune to Port-Royal was merely a sign that he was immersed in sin. Blaise, all alone in the world, loved her, was attached to her, needed her—and she would have none of that. In a letter to him demanding that he surrender her inheritance and attend her betrothal to Christ, she laid the entire weight of her father's demand that she not enter the convent on Blaise's shoulders: "It is no longer reasonable to continue my deference to others' feelings over my own. It's their turn to do some violence to their own feelings in return for the violence I did to my own inclinations during four years. It is from you [Blaise] in particular that I expect this token of friendship."[44]

Moreover, Blaise did not quite agree with her choice of Port-Royal. He had grown cool to Jansenist extremism, mainly because *they* did not accept *him*. Neither Mère Angélique nor Père Singlin had much respect for philosophers, and less for scientists. Of course, Angélique's own brother Antoine was one of the brightest young minds of the day, but he used his mental power to defend Jansenism. Anyone who wanted to peer into the mysteries of nature, as Blaise was doing, was suspect of worldliness. Moreover, Blaise had never solved the dilemma that sat between his science and Jansenist spirituality. How could the study of nature and mathematics be as great a sin as lust—indeed, be a variant of lust? Wasn't this a bit much? He objected to the level of surrender demanded by Angélique, especially her demand that he give up the intellectual life that his father had instilled in him, and that Jacqueline give up the one great gift that God had given her, her gift of poetry. But Jacqueline was in perfect agreement with Angélique. For Jacqueline, Blaise and all the world were on the side of the devil, while she and the sisters at Port-Royal alone were on the side of God. She was like the apostles, the martyrs, even like Christ himself, while Blaise was awash in the sin of Adam. If there was a sin that was committed at Port-Royal, it was the sin of spiritual pride and self-appointed martyrdom, and Jacqueline exhibited every ounce of it.

Nevertheless, out of his need, Blaise visited Jacqueline often and wrote to her often, but none of these visits seemed to solve anything. Their visits became increasingly rancorous as the conflict between them deepened. During this time, Blaise's physical condition weakened. Gilberte later described his life in this way:

> My brother, among other infirmities, could not swallow any liquids that were not warm, and even then he could only take them a drop at a time. But since he had all sorts of other maladies—dreadful headaches and severe indigestion among them—the physicians ordered that he purge himself every other day for three months. The upshot was that he had to swallow medicines, heated, drop by drop. All this resulted in a condition painful in the extreme, though my brother never uttered a word of complaint.[45]

Certainly, the feud with his sister Jacqueline, who had been his closest friend and companion most of his life, did not aid his health. Nor was it easy for Jacqueline. Several times, the novice mistress, Mère Agnès Arnauld, found Jacqueline weeping in the garden, and advised her to place her family at a bit of a distance, for they were worldly people and she had separated herself from the world. The thought that Jacqueline was expected to give up her family as a pious gesture when the Arnauld family had given up nothing of the sort smacks of a certain hypocrisy, and yet the Arnauld family was so firmly entrenched in Jansenism that they could all claim to have left the world.

Both Mère Agnès and Père Singlin tried to mediate between the Pascals but failed. It was only when Mère Angélique joined the battle that Blaise surrendered. Whenever he visited the convent he was polite, calling Angélique *"ma mère,"* and treating her with pious respect. But one day

when Blaise was visiting Jacqueline, Mère Angélique met him in the parlor reserved for visitors and lectured him on his worldliness. She admitted
that he had been faithful to what she would term "a healthy theology"
and that it was he and not Jacqueline who had brought the family to the
Jansenist movement. But as Blaise always knew, she had little respect for
worldly intellectuals, for to her the life of the mind was an indulgence
rather than a discipline. She told him that his priorities were worldly and
that he should turn them toward spiritual considerations, that Port-Royal
needed his sister's dowry, and that although the community would welcome her without it, it depended upon her contribution to help support
her. But if he was going to release her money, he should do so out of charity and not out of any other consideration. "You see, Monsieur, we have
learned from the transcendent teaching of M. de Saint-Cyran to receive
nothing for the house of God that does not come from God. Everything
that is done for some other motive than charity is not a fruit of God's
Spirit, and consequently we ought to have no interest in it."[46]

That did it. Blaise surrendered, and on June 5, 1653, he formally signed
Jacqueline's money over to Mère Angélique. The very next day, Jacqueline
made her solemn vows.

The Gambler's Ruin

Money is like muck, not good except it be spread.

—Francis Bacon, *Essays*

I'm shocked, shocked to discover that gambling is going on here!

—line spoken by Claude Rains
  in the motion picture *Casablanca*

False as dicer's oaths.

—Shakespeare, *Hamlet*

In January 1652, Blaise Pascal was twenty-nine years old, and he was alone, more alone than he ever expected to be. After the war with Jacqueline, he fell into a deep depression. He ached with loneliness, and if that wasn't enough, after he had handed over Jacqueline's portion of the family's inheritance to Port-Royal, he was riding on the edge of poverty. The Pascaline remained a toy for the wealthy and never brought in much money, and many of the debts his father owned title to were uncollectible. All the while, his dependency on Jacqueline never completely healed, for she was the last remnant of a once-supportive family. Possibly for financial reasons, or possibly out of restlessness, he moved from the family quarters on the rue de Touraine to new apartments on the rue Beaubourg. Abandoned, fidgety, empty, he quickly left the city and moved to Clermont to live with Gilberte and her family from October

1652 to May 1653, where he spent most of this time vainly trying to collect debts owed to his father's estate.

Somewhere in there, rumor has it, he met a young woman from a well-connected family and courted her. No one is sure whether this is anything more than a rumor, but there is just enough evidence to keep it tantalizing. Gilberte's daughter Marguerite said that all the while he was courting, his sister Jacqueline begged him regularly to abandon his desire for marriage altogether. She may have done this for Jansenist reasons, fearing that her brother was already too deeply embedded in the world, or she may have felt that Blaise was interested in marrying the young woman only for her money, and that his love was not genuine. In either case, it is clear that it was Pascal's intention to marry her, and this disturbed the nun of Port-Royal.

It was around this time that Blaise befriended the duc de Roannez, a man four years younger than himself who would eventually introduce him to an entirely new circle of friends. The young duke's original name was Arthus Gouffier, but he carried the title of the marquis de Boisy from his birth, and after the death of his grandfather he became the duc de Roannez. Coming from an old aristocratic family, a family that had won its spurs through military action, he was a war hero himself, having fought bravely in the battle that finally defeated the prince de Condé. Afterward, in 1651, a grateful queen and cardinal appointed him governor of Poitou, where he worked aggressively to increase the commercial and industrial development of the province. Apparently, he was a man of vision, who saw it as his duty to encourage new ideas in commerce and was therefore interested in Pascal's inventions.

It was also during this time that Pascal sent a copy of his Pascaline to Queen Christina of Sweden, who had carried on a correspondence with Galileo and had offered employment to Descartes. Unfortunately, Descartes had found that the Swedish climate did not agree with his health, and the queen's habit of calling him to attend to her at all hours reduced his health even further, until one day in 1650 he caught a chill, which led to a fever, and he died. Pascal had been encouraged to attend to the queen of Sweden as Descartes had done, for she was widely known as a fair-minded woman with a questing mind, but Pascal wisely declined, knowing that, given his health, the climate would kill him quicker than it had Descartes.

Instead, Blaise remained in France and satisfied his own need for companionship with the duc de Roannez and with the duke's sister Charlotte. Blaise had met them in Paris a number of years before, because the *hôtel* Roannez was only a short distance from the house that the Pascals occupied before leaving for Normandy. It was also to this house that Blaise and Jacqueline returned from Rouen to attend to Blaise's health and to carry on the debate about the vacuum, and Blaise may have reacquainted himself with the duke. What cemented the friendship between them, however, was a mutual interest in the intellectual currents of the day. Pascal quickly became an important figure in the duke's entourage, because he had a reputation that was all his own, earned outside the sphere of the duke's influence. He was also fairly conservative in his politics, which would have appealed to an aristocrat who feared the disorder that arose from republican thinking. What's more, Pascal's frantic desire to turn his scientific endeavors into economic gain appealed to the duke's entrepreneurial spirit, and he became ever more involved in Blaise's projects.[47]

But most of all, these two young men were attracted by a mutual desire for an authentic Christian spirituality. It was their common Catholicism, and their interest in the Catholic reform movement, that had originally drawn them together. They had encountered the spiritual wisdom of Francis de Sales, had read the books of Cardinal de Bérulle, and admired the charitable works of Vincent de Paul. As Christian gentlemen, they lived in two worlds—the world of the salon, with its witty conversations, its mildly ribald humor, and its ubiquitous gambling; and also the world of the church, where hell and damnation, salvation and eternal glory were their chief interests.

It is tempting here to think that Pascal, in his worldly period, had succumbed to the temptations of the flesh, but this is unlikely. The term "worldly" is a relative one. What would seem worldly to the people in and around Port-Royal would seem the height of religious austerity to most modern Americans. Pascal, like the young duke, was surrounded by the Augustinian God of wrath, and they strove to serve him, even love him. Like Étienne, even in his most worldly moments young Blaise was a deeply committed Catholic Christian, and it is likely that he never once had a sexual encounter in his life. Not that his health could have survived it anyway. Nevertheless, Jacqueline looked on her brother's flight into the world with increasing dismay. How much of her own responsibility for that flight she was willing to admit to remains uncertain.

In September 1653, Pascal accompanied the duke on a trip to Poitou. As governor of the region, the duke needed to oversee his estates and manage the affairs of the duchy. The region is a lush farmland, with vineyards and orchards, apple trees in long rows over the hills, and sheep sweeping white across the green, grassy knolls. Shepherds follow behind the flocks, with dogs circling around, driving strays toward the center. Once, Cardinal Richelieu had been a young bishop there. The Calvinists had once bombarded the city of Poitiers. Their commander had been Admiral de Coligny, the man whose murder had sparked the massacre of St. Bartholomew's Day.

Pascal was ill through much of the time, though he remained with the duke until early in 1654. By now, he was thirty years old, a slight man with a high forehead and a thin, almost adolescent mustache. His eyes were dark, and his nose was aquiline; his hair was long, nearly to his shoulders. His skin was pale from his constant illness; he had almost no eyebrows, and had a languor about him that belied the force of his spirit, for he was a man with a white-hot intensity, tempered by a puckish, sometimes even mean-spirited sense of humor.

But Pascal was not the only man in the duke's retinue on this visit to the country. Antoine Gombaud, the chevalier de Méré, was also along, to visit some property that he owned in the region. He had once been a knight, busy about war, but as he passed from his youth into his early forties, he tried to re-create himself as one of the new intellectuals, a man of poetry and fashion. He was fairly good at the latter but less than mediocre at the former. One of his attempted accomplishments was to try to master mathematics, something he succeeded in doing only nominally. Like some aging aristocrat out of a B movie, he had an ingratiating charm mixed with an irritating air of superiority. His one great love, however, was gambling, an aristocrat's pastime. Along with the chevalier was a younger man named Damien Mitton, a man who had arisen out of the middle classes and a fellow traveler with the *libertins érudits*. He had maintained a cool religious skepticism until his marriage, when his wife converted him into a lukewarm Catholic.

In modern language, de Méré and Mitton were the cool kids and Pascal was the nerd. Neither of them liked Pascal much—he bored them to tears—for mathematics and religion were all he talked about, and all the two of them wanted to talk about was gambling. During the long months in the duke's company, the cool kids managed to tolerate Pascal with only an occasional sneer. Meanwhile, Pascal's two sisters fretted about him from a distance. Many years later, Gilberte wrote about his journey into the world in a way that only an overbearing big sister could. Her one concern was his connection with bad companions, most notably de Méré and Mitton, and their gambling fetish. Whether or not she and her sister understood that their brother's interest in the subject was largely mathematical is uncertain.

Somewhere in there, de Méré discovered that Pascal, the "mere mathematician," had a talent that they could make use of. The chevalier had a problem. He and Mitton had been gambling nearly every night, and he had been losing money by the bucketful. At first, the chevalier had tried to bet that he could roll one six in four throws, and he had won more than he lost, but then started to lose. So he changed games and bet he could throw two sixes in twenty-four throws, but then he started to lose in a big way. One throw after another went sour, and with each bad throw he grew more philosophical. He noticed that there was an odd pattern in his luck, that he had had a slight advantage whenever he tried to roll a six in four tosses of a single die, but when he tried to roll two sixes—a "double-six"—in twenty-four tosses of two dice, he was at a slight disadvantage.[48] Why that should be so puzzled him, and worse, it was costing him money. Was he unlucky, or merely imprudent?

Then he noticed Blaise the "mere" mathematician standing nearby and put the problem to him. The question, as he put it, was how the odds would change during a series of throws. Two players agree on a game that would include a certain number of rolls of the dice. Where in this series should a player bet on getting a six? Note that the question had subtly shifted from an oracular question to a probabilistic one. In other words, the chevalier did not ask what the next throw of the dice would be, which would be something no mathematician could answer. Rather, he asked what the likelihood was of a certain outcome occurring. With that question, the idea of the future mutated, and something new was born. Pascal, who had been around gamblers and gambling most of his life, felt a vague uneasiness over the morality of the entire business, but he was intrigued by the mathematics of the question, and commenced calculating. He quickly showed the chevalier that his original observations were correct—that if he tried to bet on rolling a single six in four tosses of a single die, he was more likely to win than if he tried to go for the double-six.

Then de Méré posed Blaise a second question: two men are playing a game, whereby they agree that each man will roll the dice, and if one of them wins the roll three times in a row, he will win the bet. Each man wagers thirty-two francs, but while they are in the middle of the game something happens and they are forced to retire, after only three throws. The first man, Jean, has won twice, while his friend Jacques has won only once. How, then, should they divide the money?

Pascal answered that Jean should say, "I am certain of winning thirty-two francs even if I lose the fourth throw. Since the chance of winning the fourth throw is equal between me and Jacques, I should receive the forty-eight francs and he sixteen." But that wasn't the complete answer. What Pascal needed to do at that point was to explain how this would change if the number of throws changed. What if, for example, the game stopped after only two rolls of the dice? According to Pascal, Jean should say, "If I would have won the next toss, all sixty-four francs would have been mine. But even if I were to lose it, my share of the stake would have been forty-eight francs, as in the previous case. Therefore only sixteen francs remain to be divided by chance—eight for me and eight for Jacques—allotting me a total of fifty-six."[49]

In autumn 1653, as the golden fall leaves turned brown and fell away, the science of probability was born. The world changed utterly. In a series of letters he exchanged with his friend Pierre Fermat in Lisieux, Blaise formulated the laws of chance. From that day on, the world would no longer turn to oracles to cast a dim light on the future, for the gods had been replaced by mathematics, and risk was no longer something people suffered, but something they managed.

Letters to Fermat

As for a future life, every man must judge for himself between conflicting vague probabilities.

—CHARLES DARWIN, *Life and Letters of Charles Darwin* (1887)

The Theory of Probabilities is at bottom nothing but common sense reduced to calculus.

—PIERRE-SIMON DE LAPLACE (1749–1827)

The right of the people to be secure . . . against unreasonable searches and seizures, shall not be violated, and no warrants shall issue, but upon probable cause.

—FOURTH AMENDMENT TO THE
   CONSTITUTION OF THE UNITED STAKES

The science of probability began with a series of letters Pascal wrote to and received from his friend Pierre Fermat. Some of these letters, including the first letter that Pascal wrote, raising the question, have been lost, so we don't really know how the idea was introduced. What we do have is Fermat's reply, and this letter seems to indicate that Pascal had said something like this:

In a game of dice, a gambler bets that he will throw a six with a single die in eight tosses. The gambler throws three times and loses every time, but then for some reason the game is called off. What proportion of the stake does the gambler have the right to take with him?

Fermat's reply, written in July 1654, shows that he understood that Pascal was talking about a game that is interrupted several times, stopped and then started again, and that in each interruption the gambler is allowed to remove one-sixth of the remaining stake.

Fermat missed the point; he objected that because the stake remains intact, the probability of success in each toss is one in six, and this remains constant with each throw, so that the probability of winning the next throw is still one in six. This is where Fermat misunderstood, since what Pascal was asking was not the probability of a particular gambler winning the next toss, but the probability of that same gambler winning the game if it was carried through to its completion; he was trying to calculate the future as if it had already arrived. This required calculating the partial scores of each player, at each throw. Fermat was looking at the question in the way that the chevalier de Méré was looking at it, the way most people looked at it: as a question about dividing the spoils. But Pascal had created something new, a way of calculating the chances of a final outcome. And this was the beginning of a new science of probability. With this simple shift in focus, from the concrete game at hand to a supposed set of outcomes, the concept of risk management was born. Pascal and Fermat were not the only founders of this science, of course, but they were two of the most important. The difference between these two positions on this question—Fermat's and Pascal's—is the difference between setting the odds for the next throw, this next instance of luck, and of exploring the geometry of luck itself, of plumbing chance. With this letter, the modern world had, almost unnoticed, slipped into the murky world of partial existence.

Over the next few months, into August, the exchange of letters invented a range of new methods for calculating the distribution of points. Fermat's proposal was his *combinatorial method*, and Pascal was giddy over its possibilities, but it proved too dependent on complex calculation. It was a lot of work, and Pascal, who had invented an arithmetic machine in order to avoid tedious calculations, looked for another method, a cheat.

After all, a simple game of coin flipping was one thing, but a game of *hasard* played with two dice is much more complicated, with a far greater number of combinations. The time it would take to write them all down would make the combinatorial method impractical. What if, Pascal asked, the two players had previously agreed to keep playing until one of the players won three points? In this case, the calculation would be based not on possible outcomes but on the rules set by the two players at the beginning. Pascal considers three cases in which each player puts thirty-two pistoles into the pot:

> *Two gamblers have agreed to the game, and in the first instance gambler A quickly wins two tosses, and therefore has two points, while gambler B wins only one toss, and has one point. Either A would win with the next throw, or B would catch up with the first, and they would be tied. If the first player wins, he would walk away with the entire pot, equaling 64 pistoles. But let's say the second player wins and then the game is interrupted; then they would walk away with their original contribution, 32 pistoles. But what if for some reason they decide not to take the next throw and leave the question undecided. How would they divide the stakes then?[50]*

Pascal argues that in this case the first player should be able to say, "I'm certain to get thirty-two pistoles, for I get them even if I lose, but as for the other thirty-two, perhaps I will get them, perhaps you will get them; the odds are even. Let us divide these thirty-two pistoles in two, and you will give me one half plus the thirty-two of which I am certain."[51] So the first guy walks away with forty-eight pistoles, leaving his friend with sixteen. This is what Pascal called the *method of expectations,* which is based upon a calculation of probabilities plus an understanding of the rules of the game agreed to by the players. Here, we have an analysis of chance, but one that is made from within the context of contract law.

Pascal then moves on to another case, a variant of the first. In this case:

> *Two gamblers agreed to a game similar to the one mentioned above. In this case, gambler A has won twice, while his opponent has lost every throw.*

If they quit the game at this point, the first gambler wins everything, but if they go on to the next throw, they are in a situation similar to the first example. If the game is interrupted at that point, they will divide up the stake in the same way as they had done in the previous game, with gambler A walking away with forty-eight pistoles and gambler B walking away with sixteen. Now, gambler A can say to his friend, "If I win, I win all the money, and if I lose, I have a right to forty-eight pistoles. So give me these forty-eight pistoles that I can claim even if I lose the next throw, and let us share equally the remaining sixteen pistoles, since you have as much chance as I do of getting them." In this case, gambler A walks away with fifty-six pistoles, and his hapless friend walks away with eight and the sure but painful knowledge that he should avoid dicing with his friend gambler A in the future.

Finally, Pascal presents a third case:

Two gamblers agree on a game similar to the ones above. In this case, after the first toss, gambler A has won one throw, and gambler B has won none.

Now, let us say that gambler A wins the next throw. He will then have two throws to his opponent's none. Most people can see where this is going: gambler B is going to get hosed again. This situation would then be like the beginning of the second game, and the first gambler would have a right to walk away with fifty-six pistoles if the game were interrupted. If gambler B were to win this next throw, they would be even, and if the game were interrupted, they could each walk away with their original stake. But what if they decide not to make this next toss? In this case, gambler A could argue, "If we agree not to play, give me thirty-two pistoles, the amount I am sure to receive, and let us share equally the remainder of fifty-six pistoles." This would allow gambler A to walk away with thirty-two plus twelve (one-half of twenty-four, the difference between fifty-six and thirty-two), or forty-four, pistoles.[52]

Over the next few letters, Pascal and Fermat worked out several other methods, largely variants of those stated here. In his *Treatise on the Arithmetical Triangle*, Pascal applied these questions to work that he was doing on the arithmetic triangle, which became the basis for most of the calculations done in the insurance industry.[53] In this way, insurance companies calculate what policyholders would be due under various circumstances, making the insurance industry a very complicated dice game, with people's lives in the balance. Let us never fool ourselves that the insurance game is primarily about protection, for it is at heart a gambler's paradise, with computers.

It was Pascal's method of expectation that changed so many things, however. Pascal had essentially rewritten the question, so that what the chevalier de Méré had originally asked, which was how he could calculate the particular odds of something happening over a certain number of throws, became what each gambler could legitimately claim at each stage of the game if the game were to be interrupted. The future had been rewritten to fit the present. In short, Pascal asked, "What can I legitimately claim now, given the rules of the game and the likelihood of future outcomes?" To answer this question, he had to make the assumption that the players all agreed that once the money was in the pot, it no longer belonged to them. Thus, Pascal added a new twist to the old question of fairness. When people play games, they want them to be fair. Pascal took this seriously and added it to the nature of the game, so the games themselves implied a binding contract between the players, informal as this may be. The game, as a contract, allowed them certain expectations under certain conditions, which could permit them to walk away with portions of the pot or all of it, depending upon what happened. The players can enter the game at any time and they can leave the game at any time, and the method of expectations gives them a way to calculate what they would be entitled to at each stage of the game. Thus, Pascal's solution to the *division problem* or the *problem of points* allows them to make a decision based upon calculation. The same method applies to any decision, from dicing games to experiments in physics. How we respond to uncertainty left the oracles behind and became a question of pure mathematics from that time on. From these rough beginnings, both probability theory and decision theory began to take shape.[54]

This was a monumental change in thinking. Pascal knew this, and in his letter to the Parisian Academy, he writes about his new discovery as a geometry of chance:

Here the uncertainty of fortune is so controlled by the fairness of reason that each of the two players is always assigned exactly what belongs to him by right. This is the more to be sought by reasoning, the lesser it can be investigated by experiment. The ambiguous outcome of fortune is rightly ascribed to chance rather than natural necessity. This is why the issue has remained uncertain to this day. But if it has proved refractory to experience, it can no longer escape rational inquiry. We have turned it into an assured form of knowledge with the help of geometry whose certainty it shares. It combines mathematical demonstration with the uncertainty of chance, and having shown that these are not contraries, it borrows its name from both and proudly calls itself the Geometry of Chance.[55]

The Night of Fire

Among scientists are collectors, classifiers, and compulsive tidiers-up; many are detectives by temperament and many are explorers; some are artists and others artisans. There are poet-scientists and philosopher-scientists and even a few mystics.

—Sir Peter Brian Medawar (1915–1987)

Did you know that secret? The awful thing is that beauty is mysterious as well as terrible. God and devil are fighting there, and the battlefield is the heart of man.

—Fyodor Mikhaylovich Dostoyevsky

The mystic too full of God to speak intelligibly to the world.

—Arthur Symons (1865–1945)

Mystical experiences are what they are. Everyone has them, but most people turn them aside as moments of beauty, or as strange upwellings of happiness that comes from nowhere. Commonplace mystical experiences are too lukewarm to change lives. All too often, we turn them into psychology, into explosions of neurons in the brain, or into the working out of unfinished childhood conflicts—a lump of undigested meat, a bit of underdone potato. But mystical experiences are what they are, and they cannot be so easily parsed away. A few people have mystical experiences powerful enough to change lives. People who have had a near-death experience, for example, often come back transformed, with a new set of values that sets their lives onto a radically new course. Blaise Pascal, on a Monday night, the 23rd of November, 1654, had one of these.

Pascal had recently returned to Paris from Poitou, and in spite of all the aristocratic diversion he had enjoyed in the duke's company, his depression had only deepened. His relationship with Jacqueline was gradually healing, and yet his headlong flight into the world had not satisfied him. For the past three years, he had been living the life his father had created for him, the life of a Christian gentleman who piously lived in the world, and yet it was not enough. Of course, he kept a close contact with the duke, for the two of them had become deep friends. The duke involved him in an investment scheme, a company that he had set up to drain marshlands in Poitou, but it didn't make much money. The only good thing that happened in Pascal's life at that time was an invitation to lecture on his current work to the Parisian Academy, the grand institution that Père Mersenne's little seminar had mutated into.

Something was moving inside of Pascal, a dark night of the soul leading to a conversion. He was unhappy with his life, though it was the life he had always lived. In spite of his previous rancor with the sisters of Port-Royal, he consistently noticed on his visits to see Jacqueline and his nieces, Gilberte and Florin's daughters, who were living at Port-Royal, that his sister possessed a serenity that he could not find in his own life. He had lived a life of the mind, and the life of the mind had not filled him up. His emotional life was stifling, and what he needed was a change of heart. He was disgusted with the world and with himself.

Just before or just after the night of his great conversion, Blaise Pascal wrote a short piece entitled "On the Conversion of the Sinner." Before the conversion, he said, there is a deep dissatisfaction with the world: "The soul can no longer serenely enjoy the things that captivated it. Constant scruples attack the soul in its pleasure, and because of this introspection it no longer finds the usual sweetness in the things to which it once abandoned itself blithely with an overflowing heart."[56]

This was a perfect description of what Pascal's own life had become. This dissatisfaction was part of the "first stirrings" that God was causing in the sinner's heart. "But the soul finds more bitterness in the disciplines of holiness than in the futilities of the world. On the one hand, the presence of visible things seems more powerful than the hope of the things unseen; on the other hand, the permanence of things unseen moves it more than does the frivolity of visible things. And thus, the presence of the one and the constancy of the other fight for the soul's affection; the emptiness of the one and the absence of the other awaken its disgust."[57]

This apparent contradiction in the heart of the sinner produces great confusion and discord. But this is only temporary, for eventually the sinner begins to realize that visible things are already passing, even in their enjoyment. "The soul considers mortal things as already dying and even as already dead. It is terrified by this realization, this certainty of the annihilation of all that it loves as the ticking by of each moment snatches away the pleasure at hand; it is terrified when all that is dearest slips away into nothingness at every moment, and by the certainty that there will come a day when all the sweet things of this earth will be gone, and the soul will be destitute of the things it placed its trust in."[58]

A knowledge too terrible to contemplate, a truth of life too fearful to hold—this is what drove Pascal, bit by bit, from the world. This was a knowledge that he had already known as a set of platitudes, something one says at funerals, but not something one takes to heart. It is interesting to speculate how these reflections had come upon Pascal as if newly minted after he had spent a greater part of the year in the company of gamblers. Perhaps it was the chevalier de Méré and his questions, so full of the uncertainty of the future, so full of the desire to hedge his bets, that started Pascal upon this road.

Once back in Paris, Blaise changed his residence once again, moving from the rue Beaubourg to the rue des Francs-Bourgeois, not far from the house that his father, Étienne, had brought the family to when they had moved to Paris from Clermont, but, more important, near the convent of Port-Royal de Paris. It was only a short walk away, a matter of ten minutes, past the queen's basilica of the Val-de-Grâce, the very church where years before Cardinal Richelieu's agents had caught Anne of Austria in the act of sending treasonous letters to her brother, the king of Spain. In these new quarters, Blaise could visit his sister anytime he wished, and did so regularly.

Likely, Port-Royal's parlor was similar to the ones used by Discalced Carmelites—a poorly lighted room with hard-backed chairs gathered near a wide grille or set of bars cut into a wall. On the other side, there would be a curtain that could be opened or closed whenever the sisters were present. The old-fashioned grille work, which is most likely what existed at Port-Royal at the time, was more like a semiopaque screen, so that the sisters would appear almost ghostly on the other side, like disembodied voices speaking out of shadows. Nevertheless, Blaise could visit his sister often, so often that Jacqueline wrote to Gilberte that "it would take a volume to tell you about all the visits one by one. This was such that I sometimes thought I had no time for any other work."[59] In these long sessions, Blaise opened his heart to her, and the rancor of all those months at the time of her profession was gone. She pitied him deeply, for in the midst of all of his busy life and all of his success, he hated his own life. He was living the very existence he described in his short tract on the conversion of the sinner; he was the sinner described on those pages, and what appeared as a theological discourse was actually a most honest confession. And yet he felt no attraction to a God who would ask him to give up the acuity of his mind. This was his stumbling block, for although he could easily give up the pleasures of the flesh—good food, good conversation, a host of entertainment and diversions—he did not wish to give up his life as a philosopher and scientist. His own body had caused him nothing but pain for twenty years. What were the pleasures of the flesh, when the flesh had been so full of pain? But how could he give up the one thing he relied upon for the truth—his own reason, his own intellect?

Sadly, had he turned to the Jesuits for spiritual guidance, they would not have required such a sacrifice, for it was their stated belief that one could find God in the midst of all things, that finding God was not a matter of abandoning the world as the Desert Fathers had done, but of choosing Christ in the midst of one's life in the world. One could live in the world, one could live the life of a Christian gentleman, and still find God there.

But the Jesuits' was not the path he had taken. Blaise had always been attracted to a more rigorous theology. For Jacqueline and the sisters of Port-Royal, it was not enough to merely assent to God in the mind; it was necessary to love God in the heart. But one could not love God and any other thing. Oddly, this is where Jansenism was both right and wrong, for though one of the two great commandments that Jesus held to was that we should love God with our whole heart, with our whole soul, with our whole mind, and with our whole strength—which sounds an awful lot like loving God, and only God—the other great commandment was that we should love our neighbor as we love ourselves, so that self-love was not anathema, but a component of the love of God. Perhaps, however, Christians could truly understand this connection only after the invention of psychology.

For months, Jacqueline and Gilberte conspired to pray for their brother, like St. Monica praying for her son Augustine, in the hope that he would find spiritual peace. Jacqueline understood that she owed a great debt to Blaise for her conversion and her life at Port-Royal. Then, a few months later, on Monday, the 23rd of November, 1654, Blaise met God. He never told anyone about it, never mentioned it, never said a word, and only wrote a short memorial about the experience and pinned it to the inside of his clothing, near his heart. It was only after his death, nine years later, when Blaise's nephew was going through his clothing, that they found it. A servant felt through the garment and found what he thought was extra padding stuffed into the doublet; on further examination he found a piece of crumpled parchment, with a faded piece of paper wrapped inside. And there they found, in Blaise's own handwriting, the story of his "night of fire."

The year of grace 1654,

Monday, 23 November, the feast of St. Clement, pope and martyr, and of others in the martyrology.

The Vigil of St. Chrysogonus, martyr, and others.

From about half past ten at night until about half past midnight,

FIRE.

GOD of Abraham, GOD of Isaac, GOD of Jacob
not the God of the philosophers and of the learned.
Certitude. Certitude. Feeling. Joy. Peace.
GOD of Jesus Christ.
Deum meum et Deum vostrum. [My God and your God.]
Your GOD will be my God.
Forgetfulness of the world and of everything, except GOD.
He can only be found by the ways taught in the Gospel.
Grandeur of the human soul.
Righteous Father, the world has not known you, but I have known you.
Joy, joy, joy, tears of joy.
I have departed from him:
Dereliquerunt me fontem aquae vivae. [They have forsaken me, the fount of living water.]
My God, will you leave me?
Let me not be separated from him forever.
This is eternal life, that they know you, the one true God, and the one that you sent, Jesus Christ.
Jesus Christ.
Jesus Christ.
I left him; I fled him, renounced, crucified.
Let me never be separated from him.
He is only kept securely by the ways taught in the Gospel:
Renunciation, total and sweet.
Complete submission to Jesus Christ and to my director.
Eternally in joy for a day's exercise on the earth.
Non obliviscar sermones suos. [May I never forget his words.] Amen.[60]

So Jolly a Penitent

The weakness of little children's limbs is innocent,
not their souls.

—St. Augustine

To Carthage I came, when all about me resounded a
cauldron of dissolute loves.

—St. Augustine

Give me chastity and continence, but not just now.

—St. Augustine

God of Clothilda, if you grant me a victory I shall
become a Christian.

—Clovis (466–511)

Suddenly, Monsieur Pascal was a changed man, and no one knew why. Everyone who knew him marveled at the change: one day he was drowning in confusion, and the next he was free of it. No one knew anything about his experience on that night in November, his night of fire, and would not know about it until after his death, so they could only speculate. Of course, there were stories to explain his conversion, so that even after he had died and the family found the memorial pinned to his doublet, they simply dusted off the old stories and shifted them to explain the cause of his mystical experience. By this time, the stories were well trod, and the fact that they were speculative didn't mean that they were devoid of truth.

One story tells how Pascal had nearly died in a carriage accident on a day when he was out with some friends; the driver lost control of the horses while they were crossing a bridge over the river Seine, and the carriage and passengers nearly ended up in the river. Supposedly, this accident put a certain rush on his desire for conversion, something he might have put off until another day had he not been faced with the possibility of sudden death. The other story is less dramatic, the one told by Marguerite Perier about her uncle after his death—that he attended Mass one Sunday and heard a particularly poignant sermon given by Père Singlin, one that forced him to begin thinking about his life, one that led him to a sudden change of heart, perhaps even led him to the "night of fire." But this seems a bit too much like a pious set piece, as do many of the stories told about Pascal by the Perier family. And Pascal must have heard many powerful sermons before that date, so why did that one affect him so? The story feels like a literary type, for too many saints have had conversions of heart after listening to sermons full of hellfire. Supposedly, St. Anthony of Egypt heard such a sermon and, picking up the challenge, sold everything he had, left the city of Alexandria, and went to live in the desert. The same thing happened with St. Augustine by a chance reading of the Bible. There is a theory that the hand of God touches people through sermons and bits of Scripture blown open by sudden gusts of wind. And even though the theory is not unreasonable, it has been told enough times to be suspicious, or at least clichéd.

Either way, Blaise Pascal was a changed man, and he informed his sister Jacqueline that he was ready to place himself at the disposal of the Augustinian movement, to turn his back upon the world. One can only imagine how thrilled Jacqueline was at hearing this, though she may have been puzzled at first, for his change of heart was sudden indeed.

Eventually, Blaise ended up under the direction of Père Singlin, though the good father, while exceptionally holy, was not a strong enough intellect to direct the younger Pascal, and perhaps was wise enough to know it. Therefore, he sent him on to the country monastery, for "M. Singlin believed when he saw this great genius that it would be well to send him to Port-Royal-des-Champs, where M. Arnauld would lead him in what pertained to the higher sciences, and where M. de Saci would teach him to despise them."[61]

But as Pascal had written in his memorial, unknown to everyone else, he had already decided to give "total submission to Jesus Christ and to my director," and thus to walk the Jansenist path, for he had found the God of Abraham, Isaac, and Jacob, and not the God of the intellectuals. On the other hand, his life hadn't changed much outwardly, and this troubled Jacqueline, for she had no idea of his inner life and could judge only by what she saw. Though he had obviously found a measure of peace, he still lived in the same residence in the Faubourg Saint-Michel, and still kept the same staff of servants. The woman who ran his household was the sister of Étienne's old housekeeper, Louise Default, the woman who had taken care of the children when Étienne had to leave town to avoid facing the wrath of Cardinal Richelieu, and who had since run the family home for many years. This sister had her husband, and also her two daughters, all of whom worked in Blaise's service. Besides that, the younger Pascal had hired a cook to fix his meals and a footman to help him travel about town. His house, near the church and across the street from one of the great *hôtels*, had a garden, with flowers and fruit trees, where Pascal could sit on a summer's eve and listen to the birds and the ringing of the bells.

All of this concerned Jacqueline, for Blaise did not seem to have changed his habits in the slightest, though Gilberte noted that he had begun to rely less and less on his servants, and even made his own bed. He was a gentleman, however, and lived a gentleman's life. Like his father, Blaise was an *honnête homme*, and was by all appearances still a man of the world. So how could such a one, "so jolly a penitent,"[62] claim that he was ready to give up life in the world when the world was so much with him? He had not even broken his association with that clutch of notorious gamblers the chevalier de Méré and Monsieur Mitton, who hadn't changed their ways one jot or tittle, nor showed any inkling to do so. She was more comforted by the presence of the duke, however, for he, too, seemed to be interested in the spiritual life, and though he was not willing to change too much, nor run headlong toward Port-Royal, he was certainly ready to become a fellow traveler with the Jansenist movement.

Had Jacqueline known the complete story of her brother's mystical experience, she might have had very different feelings about his professed change of heart; she might have realized that this jolly penitent was basing his changes on something more profound than anyone in her circle had ever experienced. What Blaise had learned from his night of fire was that there is a kind of certitude that has little to do with proper methodology—"certitude, certitude, feeling, joy, peace"—the surety of direct experience. In a world where doubt had become intellectually fashionable, where theological ideas once bolstered by cosmology had been overturned, Blaise had learned that one could become certain of God's existence and of God's nature by direct encounter, bypassing reason altogether. While he maintained a desire to continue his work as a mathematician and philosopher, he cancelled his running correspondence with Fermat on probability, ceased tinkering with his arithmetic machine, and postponed the distribution of his treatise on the arithmetic triangle, though it was already printed. He began to question the power of reason itself; while never really doubting its capacity to reveal truth, he decided that that capacity was limited to lesser truths and could not supplant the truths of revelation. Piety was no longer an empty practice, and reason was no longer a royal road to truth.

But Blaise's newfound Jansenism had one stumbling block: he was ready for them, but were they ready for him? Père Singlin and Mère Angélique were still suspicious of worldly philosophers, for they both believed that the desire to puzzle out the world meant that one still belonged to the world. So how to break him from this addiction? The only answer was to send him on a retreat to Port-Royal des Champs, where he could take time with the solitaries, *les messieurs des granges,* "the gentlemen of the barns," as the locals called them, to pray and to receive instruction. The solitaries of Port-Royal were a group of laymen who had left the world to live a life of penance without actually joining a religious order. Back in 1625, Mère Angélique Arnauld had brought her troupe of sisters into Paris from the country because the old monastery of Port-Royal des Champs had grown dilapidated through the centuries. Once in Paris, the community flourished both in size and in wealth.

A number of important people soon joined the movement, most notably the son of Catherine Arnauld, the oldest sister of Mère Angélique, who had married a man by the name of Lemaître. Their marriage didn't last very long, but it did produce one son by the name of Antoine, born in 1608. When the marriage broke up, Catherine returned with her son to the Arnauld family home, and the boy was raised in the home of that fiercely loyal clan, so loyal that they followed one another into Jansenism. He grew up to become one of the brightest young legal minds of his day—eloquent, intelligent, and charming in the extreme—and everyone at the royal court expected him to have a career among the planets. But then he met the abbé de Saint-Cyran and, with half the other members of his family, decided to retire from the world and live a life of penance. Because of his prominence, his conversion was anything but private, and the loss of a man of such talent stirred the rancor of Cardinal Richelieu, the king's first minister. The cardinal knew that he couldn't afford to lose too many men like that, especially to the radicals of Port-Royal. Then the *Augustinus,* the originating work of Jansenism by Cornelis Jansen, was published in 1640, and soon after, in 1643, Antoine Arnauld published his notorious tract on frequent Communion, *De la fréquente communion.* The cardinal's irritation with Port-Royal became deadly. The famous Antoine Lemaître then joined the fray by writing a pamphlet condemning "laxity" and hinting that the laxists included some of the most important churchmen in the nation, and most especially the Jesuits. Richelieu was furious, and set his eye on them.

At first, Lemaître lived in a small house within the compound of Port-Royal de Paris, but after Cardinal Richelieu had the abbé de Saint-Cyran arrested, Lemaître quietly left town and took up residence in the old monastery of Port-Royal des Champs. Here, other young men of similar status and spiritual concerns gathered with him, and here they worked on improving the old monastery, replanting the gardens and repairing the old buildings. Over the years, the solitaries managed through consistent hard work to breathe new life into the dying monastery. They drained the swamps and planted new fields, and with time the place began to become beautiful once again.

Soon after the abbé de Saint-Cyran had been thrown into prison and Antoine Lemaître had retired to the country, two of Antoine's brothers—each sporting one of the odd faux titles of aristocracy that the Arnauld family seemed to enjoy, one Simon Lemaître de Séricourt and his brother Isaac Lemaître de Saci, otherwise known as "de Saci"—joined him in the country. The younger brother, Isaac, had decided to study for the priesthood, and when he was ordained, he and Père Singlin split the duties of spiritual direction between them, with Père Singlin taking the Port-Royal community in Paris and Père de Saci taking the gentlemen of the barns in the country. Together with the solitaries, he translated many of the works of the church fathers into contemporary French and, along with Pierre Nicole, founded the *petites écoles,* a school for young boys, a school that eventually produced some of the more prominent minds in France, men like the poet and dramatist Jean Racine and the historian Louis-Sébastien Le Nain de Tillemont.

In modern terms, Lemaître de Saci would have been described as "laid-back"—quiet, plodding, with little obvious fire. His great talent was to be able to talk to anyone, and to adjust his style of direction to fit each new directee. By all accounts, he was a fine listener and a natural-born psychologist. To carpenters, he talked of carpentry. To painters, he talked of painting. To philosophers, he talked of philosophy, and thus led them through the world of their making into the world of God's making. Much of what he and Blaise talked about in those sessions during Blaise's retreat and afterward has been lost to us—all except what was said in one session, which had been witnessed by de Saci's personal secretary, Nicholas Fontaine, who took copious notes during the session and wrote them down years later in a pamphlet entitled *Entretien avec M. de Saci.*

By this time, Blaise was something of a celebrity among the European intellectuals. His adding machine, the Pascaline; his work on the vacuum; and, most recently, his letters with Fermat on probability had spread his name across the continent, and he came to Père de Saci as one of the great men of his time. De Saci, however, was not all that impressed, and though he recognized Blaise's obvious talents, he was concerned about falling prey to the glamour of Pascal's intellect. Like Mère Angélique and Père Singlin, he was too deeply entrenched in the thinking of St. Augustine to be impressed by worldly success. He began the conversation by asking Pascal what books he had been reading, and of course Pascal responded by expounding on two works of philosophy. He had been reading, he said, the works of Epictetus the Stoic and the short essays of Montaigne, and found them both instructive.

What followed was an encounter between two basic worldviews. Pascal wanted to find a way to integrate his new faith into his previous scientific work. He had long believed that the man of science ought to pay attention to what he sees rather than slavishly follow some already determined philosophical system, as Descartes had done. Descartes, the philosopher, had tried to find a new way to build an Aristotelian worldview in a time of skepticism. Pascal had always argued that the discovery of the existence of the vacuum and other scientific achievements should be based upon proper methodology and upon the witness of one's own eyes rather than through faithfulness to a philosophical tradition. Shouldn't this same attitude be brought to bear on the spiritual life? This made perfect sense to Pascal, for his newfound spirituality came out of a direct experience of the God of Abraham, Isaac, and Jacob, and not out of some philosophical insight.

Pascal then began to discuss the works of Epictetus with de Saci, saying that Epictetus was a philosopher of duty, and that his wisdom, pagan though it was, often rang true with the teachings of Augustine. Epictetus taught that everything belongs to God and that all the things we enjoy are only on loan, Pascal told him, summing up the Stoic ideal. "Never say, says he, 'I have lost that.' Say instead, 'I have given it back. My son is dead. I have given him back. My wife is dead. I have given her back.'"[63]

But then he went on to criticize the same ideal by saying that Epictetus was wrong in assuming that people have the power to do good and that people are truly free to do their duty. Epictetus did not understand the fallen nature of humanity and therefore in his paganism could believe that an act of suicide could be a call from God under a condition of persecution.

After this, Pascal discussed his favorite author, Michel de Montaigne, whom he preferred over Epictetus because he was a Christian and in various places ably defended the Christian faith. He had used the skepticism of the day to fight the *libertins,* and "since he wished to discover what morality reason ought to prescribe independent of the light of faith, he based his principle upon that supposition. . . . He places all things in a universal and so general doubt that this very doubt carries itself away."[64] And so the doubter doubts even his reason and his own quest for knowledge, and therefore leads himself into a perpetual circle, a strange loop of doubt.

De Saci's response was, following Augustine, that such philosophers could lead one astray, into the swamp of intellectual pride—a dangerous pleasure, a *iucundidate pestifere,* where the great mind thanks God for forgiveness while still enjoying the vanity of the world. But what is the use of such readings, if they are so very dangerous?

Pascal responded that Epictetus had the remarkable ability to disturb the complacency of those addicted to material pleasure by showing them that they were slaves to their own flesh, which eventually must die. But while Epictetus can prick us from our sleep, he can lead to pride, the pride that says that we fallen humans have the power to save ourselves, when only God has such a power. Montaigne, on the other hand, is wonderful for attacking the pride of the narrow-minded, those fools who think they can find truth outside the faith through science. But Montaigne, for all that, leads to despair, the doldrums of the intellect, which, without faith, merely bobs up and down in the water, without a breeze to carry it forth.

By the end of the conversation, the two men, priest and mathematician, had demonstrated to each other that they were perfect agreement. Pascal had finally won his spurs in the Augustinian movement, and quickly moved into the inner circle.

The Jesuit Menace

*Therefore you see, Fathers, that mockery is sometimes more
suited to making men abandon their aberrations, in which case
it is an act of justice; because, as Jeremiah says, the actions of
those who err "are vanity, the work of errors: in the time of their
visitation they shall perish: vana sunt et risu digna."*

—BLAISE PASCAL, *Provincial Letters*

*Therefore, let them consider, before God, to what an extent
the moral code spread abroad everywhere by your casuists is
shameful and pernicious to the Church; to what extent the laxity
of behavior they are introducing is shocking and immoderate.*

—BLAISE PASCAL, *Provincial Letters*

Père Vincent de Paul was not happy with this new crop of Augustinians. He'd once been friendly with the abbé de Saint-Cyran, and admired his religious sensibilities, his sense of discipline, and the depth of his commitment, but he was less impressed with Saint-Cyran's heirs at Port-Royal. The Arnauld family was particularly contentious, and Antoine's short tract on frequent Communion infuriated the saint no end. For Père Vincent, the Eucharist was central to Christian spirituality, the food of the saints and God's great aid to salvation. Anyone who kept God's people away from Communion, even while proclaiming the terrible holiness of the sacrament, could not be representing God's will. Moreover, he knew that the people at court were divided over the Jansenist movement. Some of the most powerful courtiers were partial to the sisters at Port-Royal, while the two cardinals—first Cardinal Richelieu, and then the less powerful, though certainly adept, Cardinal Mazarin—had little use for them. As Queen Anne's spiritual adviser, Vincent de Paul expressed his views openly, but the saint was not alone in his concerns. Jean-Jacques Olier, the holy founder of the great seminary at Saint-Sulpice, had joined him in his protest, and nothing less than the powerful Society of Jesus backed their play.

When Antoine Arnauld published his short tract on frequent Communion in 1643, with the encouragement of the abbé de Saint-Cyran, the publication had ignited a theological firestorm. Some were sure that whenever a sinner received Communion, the sinner was polluting the sacrament, insulting God himself. Others were just as sure that the sacrament had been given to human beings as a cure for sin, not as a reward for perfection, and that postponing its reception in a misguided search for penitential perfection was a terrible risk to the soul. At the urging of the Jesuits, the pope, Urban VIII, who had tried Galileo, reviewed the Jansenist question and published a papal bull, *In eminenti,* giving the Jansenists a mild rebuke but never mentioning them by name. In 1644, Vincent de Paul encouraged the queen to ask Arnauld to take his notions before the pope, argue them there, and let him decide, and Arnauld agreed to do so, swearing his fidelity to the church and protesting his desire to remain a Catholic. He was all set to go when Urban VIII died. It took months for a successor to be elected, and so the issue was tabled.

But it did not go away. Too many Catholics saw the rigorous Jansenists as departing from the true spirit of Christianity, for they seemed more interested in sin than in salvation. Besides, for many churchmen there was an irritating single-mindedness to the Jansenist version of the faith, one that bordered on fanaticism—a single-mindedness that was as much a product of Arnauld's personality as it was of Jansen's theology. Arnauld had been something of a logic missile at the Sorbonne, driving his ideas and his principles home with a precision and a fearsome ruthlessness. He was more than aware of, and embarrassingly honest about, the hypocrisy of the ruling classes in France, how they preached Christianity to the poor but then acted as if Christian principles didn't apply to them when their own drive for power was at stake. For Arnauld, Christian principles, as interpreted by Augustine, always trumped the world's corrupt logic.

Cardinal Richelieu had been furious over Arnauld's pamphlet, but he had already thrown the abbé de Saint-Cyran into prison. There was little else he could do except hang the man, but the abbot had too many powerful friends for that. Besides, it wouldn't look good for one prince of the church to hang another. Disturbs the good people at their breakfast. After Richelieu died in 1642 and Mazarin took over, things seemed to get better for the Jansenists, because Cardinal Mazarin was less interested in matters of religion than his predecessor. But that didn't last long, because there was no such thing as even the smallest crack of a separation between church and state; troubles that affected the church immediately affected the state, and vice versa. No one had forgotten the horrors of the religious wars, and no one wanted them to come back. The answer was to build a united front between church and state. The dangers of theological disputes throughout Europe had been etched on people's souls. In France, therefore, there was one king, one faith, one law.

By 1655, Antoine Arnauld had become the premier spokesman for the Jansenist movement. He had aged into a short, thick, balding man with a penetrating gaze and the manner of a badger, for he had never lost the relentless fire he had possessed at the Sorbonne. Those who disagreed with him were more than mistaken; they were heretics. Seemingly, there were plenty of those, because the Jansenist movement had not spread much in France and was limited to a small number of converts, and a good portion of them were Antoine's relatives. His older brother, Henri, was the bishop of Angers, and he was a violently outspoken Jansenist. Many of the other converts, however, were members of the upper class—dukes and countesses, a part of the old aristocracy—joined by a clutch of middle-class lawyers and bureaucrats who had decided to abandon their portfolios to save their souls. The ordained clergy who belonged to the movement were all diocesan priests and not members of religious orders, and so much of the pressure against the Jansenists came from the religious orders, especially the Jesuits.

Though they would not admit it, the Jesuits disagreed with Saint Augustine and his theory of original sin. They were Molinists to a man, and believed that human beings were radically free, though wounded in their ability to exercise that freedom. God's knowledge of the future, they argued, is not absolute, and they posited a "middle knowledge," a *scientia media,* whereby God knows how any rational person would act under any condition. While God knows what people are likely to choose, that knowledge does not determine what they do choose. Therefore, human beings share in, and in some small way limit, the power of God, for by creating humanity in his own image and likeness, God bestowed upon them the power to bring new things into the world through their freedom. God's grace does not subvert human freedom, but acts with it. It is not "efficacious," in the sense that it does not force the person who receives it to convert, but aids them in their free choice. It is therefore "sufficient" in the sense that it is enough to affect a change when working in concord with human free will. In a very real way, this is the position that has survived, the one that has come down to us in our own time. It is the position that may well be the root of the modern idea of liberty, for how can freedom from government control mean anything if people are not metaphysically capable of free acts?

The Molinists opposed Augustinians of all types, especially the disciples of Michael Baius and Cornelis Jansen. Following the ideas of Augustine, the Augustinians believed that after the Fall, human will was so polluted that human beings could not choose to do good but could choose only to do evil, for sin had become an essential part of human nature, called concupiscence. Those whom God saved were saved by "efficacious grace," grace that always carried out its effect, regardless of the will of the recipient. Those who are saved, therefore, are saved in spite of themselves.

The Molinists, by which we can read the Jesuits, accused the Jansenists of being crypto-Calvinists. What was really happening, however, was that the old war that had taken place between Augustine and Pelagius over human freedom was being revisited, and the church, by siding with the Jesuits, was slyly condemning one of its greatest theologians. To be sure, theologians of all stripes had chafed for a thousand years at the muscular theology of redemption that Augustine had set down. During the high Middle Ages, they had invented such notions as purgatory and limbo to soften Augustine's blow. Augustine had declared that all the unbaptized, even the innocent, would be burning forever in hell. He also proclaimed that all the imperfect, those who had failed to cut out the final traces of sin through the grace of God, would be burning right beside them. He also believed that even the righteous who were culled from the elect would burn along with the others, to pay for the sin of Adam. This was too harsh, even for the Middle Ages.

The war was joined in earnest in 1649, when Nicolas Cornet, the widely respected syndic, or censor, for the Sorbonne, appeared before the theology faculty and presented them with seven propositions for condemnation. The first five of these were drawn from the *Augustinus* by Cornelis Jansen, while the last two were drawn from Antoine Arnauld's little tract on frequent Communion. By this time, Cornet was a seventy-year-old man, and was well known to both Cardinal Richelieu and Cardinal Mazarin. Whether Mazarin had anything to do with Cornet's proposals is uncertain, but the content was taken directly out of the Jesuit playbook, for these five propositions were at the heart of the conflict, and the Jesuits would have known that better than anyone. The very next year, an assembly of the French clergy gathered and offered their support of Cornet's condemnation of the *Augustinus,* but refused to do likewise for the two propositions taken from Antoine Arnauld's tract. Perhaps this was because Jansen was already dead, whereas Arnauld was very much alive. And writing.

The war gathered force as four French bishops sent a letter to Rome asking for the pope's condemnation, while at the same time eleven other French bishops met at Port-Royal and sent another letter to Rome asking for the pope's approval. From the beginning, the popes were wary about getting involved in French politics, for that was a murky swamp that could trap the unseasoned traveler. Urban VIII's weak condemnation of the Jansenists set the pace for the papal response, and when all of these letters attacking and defending the theology of Port-Royal landed on his desk, Pope Innocent X had no desire to spend political capital on a theological battle that seemed to have no end and that was so subtle that even professional theologians were confused. However, France was one of the most important Catholic nations, and the French church was a church he could not ignore. He knew how touchy the French were about their "Gallican liberties" and their fear of papal interference. But at that moment, all hell was breaking loose in the first daughter of the church, for the people were rising up against their queen and cardinal, and, after the Reformation and the following wars of religion, the pope knew he needed to act, but to act very carefully.

In April 1651, he gathered five cardinals and appointed them to a committee to study the question. The Cardinals then assembled another staff of thirteen theologians and lawyers to assist them. Over the next two years, this group, the Assembly of the Clergy, met fifty-one times. Lobbyists from all over Europe, representing every point of view and every small constituency, gave fiery speeches and then met privately with one or two of the cardinals and offered favors. On May 31, 1653, the committee produced an ecclesiastical constitution entitled *Cum occasione: Errors Said to Have Been Extracted from the "Augustinus" of Cornelius Jansen,* siding with the faculty of the Sorbonne and condemning the first five propositions that Cornet had taken from the *Augustinus.* In high ecclesiastical language, the five propositions read like this:

1. Some of God's precepts are impossible to the righteous, who desire them and endeavor to keep them, according to the present powers which they have because the grace by which they are made possible is wanting.

   Declared and condemned as rash, impious, blasphemous, anathema, and heretical.

2. In the state of fallen nature one never resists interior grace.

   Declared and condemned as heretical.

3. In order to win or lose merit in the state of fallen nature, it is not required that men be free from necessity, but freedom from external compulsion is sufficient.

   Declared and condemned as heretical.

4. The Semipelagians admitted the necessity of a prevenient interior grace for each act, even for the beginning of faith; in this they were heretics, because they desired to define this grace as one that the human will could either resist or obey.

   Declared and condemned as false and heretical.

5. It is Semipelagian to say that Christ died or shed his blood for all men without exception.

   Declared and condemned as false, rash, scandalous, and understood in this sense, that Christ died for the salvation of the predestined only—such beliefs are impious, blasphemous, contumelious, dishonoring to divine piety, and heretical.

The first part of each of the five points is a statement taken from the *Augustinus*—that is, something that the Jansenists held to be true—while underneath it is the official church condemnation of that very statement, calling it heretical and so forth.[65] The gist of the five points in contention is that Augustinians denied human beings the power of full moral agency. People could commit evil on their own but not good, for doing good requires a special "efficacious" grace from God, and, once given, that grace could not be denied. With it, one could not do evil; without it, one could not do good. The question was whether people were puppets in the hand of an all-powerful God, thereby making God's power absolute, or whether they were moral agents capable of free actions, thus in some small way limiting the power of God.

Antoine Arnauld set off for Rome to plead his case. While doing this, he came up with an almost Jesuitical distinction that would keep the Jansenist issue alive for centuries. In order to keep his Catholic standing,

Arnauld declared that the statements in the pope's Apostolic Constitution were correct in every part, but that Jansen had not actually held them. He acknowledged the right of the pope to teach in matters of law and morality (*droit*), which every Catholic acknowledged as the pope 's right as the Vicar of Christ. However, the pope did not have the power to teach on questions of fact (*fait*). This meant that the pope was right and just in proclaiming that those ideas embodied in the five points in the ecclesiastical constitution were morally and theologically unjust, and that anyone who held those ideas would be a heretic. However, the pope had no authority when he spoke on questions of fact. If the pope said that Rouen was the capital of France, no one had to believe it because of papal authority. Likewise, if the pope said that Jansen had actually taught the ideas embodied in those five points, then the average Catholic would not have to pay attention to this, for it would be a question of fact and not of faith and morals. Arnauld essentially lobbed the ball back into his enemy's court by claiming that Jansen had been falsely and maliciously accused of teaching heresy when he was doing no such thing, and that it was his accusers themselves who were the heretics.

Nobody but Jansenists bought this. Though this argument was a good rallying point for the Augustinians, it didn't change the political climate back home. Very quickly, in March 1654, the Assembly of the Clergy, the representatives of all the clergy in France, fired the ball back to Arnauld by claiming that the pope's Apostolic Constitution did indeed condemn Jansen in matters of fact as well as in matters of faith and morals, that Jansen did say these things, and that these things were indeed impious, blasphemous, contumelious, dishonoring to divine piety, and heretical. So there!

From that point on, things just got worse and worse for the supporters of Port-Royal. On January 31, 1655, Blaise Pascal had just finished his retreat and returned to Paris when the storm broke. It was on a Saturday, and the duc de Liancourt, an outspoken Jansenist supporter, arrived at his parish church, a church that was under the control of the priests of Saint-Sulpice, to make his confession. But when he arrived, his parish priest, instead of giving him absolution, gave him a short lecture on his heretical associations with the gentlemen of Port-Royal. The priest denied him absolution, and then informed him that he would also be denied the Eucharist if he came to the Communion rail. This decision had come down from on high in his order, and had been made by the pastor of the parish, Jean-Jacques Olier.

Antoine Arnauld sat down at once and fired off another pamphlet, *Lettre d'un docteur à une personne de condition (Letter of a Doctor of Theology to a Person of Rank)*. Obviously, he said, the followers of Augustine were loyal children of the church, and they had conformed themselves willingly to the Apostolic Constitution, and he joined the church in condemning them. However—here was the sticking place—the pope was simply misinformed about the facts of the matter. No churchman needs to follow the pope when the pope speaks on matters of fact, and therefore the Augustinians were loyal Catholics because they condemned the five propositions along with everyone else even while they did not believe that the five propositions applied to them.

"Nonsense!" said their opponents. "The Jansenists are heretics," said François Annat, the Jesuits' spiritual director to the king. "They are Calvinists, all of them, for they have adopted the Calvinist theory of grace, because they believe that interior grace, which is necessary for people to choose the good, is given to some and not to others, and that some people sin because God wills it so, because God did not give them the grace to do otherwise."

Then Arnauld made a mistake. He quickly sat down and wrote a book, *Seconde lettre à un duc et pair,* two hundred pages long, where he stated his case in meticulous detail, using an old chestnut out of the Jansenists' playbook that immediately set everyone's hair on fire: St. Peter denied Jesus as Our Lord was being taken off the cross, and he did so because the "interior grace" was denied him. Therefore, he denied Jesus not by his own choice but by the hand of God. Hang on, Arnauld's enemies said. Peter denied Jesus because of his cowardice. We say that this was his choice, not God's. You say it was God's because God had denied him the grace. But that is what the Calvinists teach, which makes you closet Calvinists and not Catholics, for you deny all free will, which leaves only predestination.

Now the battle was joined. The French bishops finally acted in May, under the influence of Mazarin, of course, and wrote up a formulary that stated that the five propositions condemned by the Apostolic Constitution were indeed heretical and that they were indeed found in Jansen's theology—no ifs, no ands, no buts. Accusations of heresy arced over Paris like missiles over the North Pole, and everyone was in deadly earnest, for the loser could well be hanged or find himself running for his life to the Netherlands or, worse, England! All the priests, both diocesan and religious, had to sign the formulary, a statement of approved belief, and swear to it. But this wasn't enough. In November, the faculty of the Sorbonne denounced Arnauld's *Seconde lettre*. Not to be outdone by the pope, the faculty set up a commission of their own, this time with six doctors of theology, to study the questions both of fact and of law. Were those propositions indeed in Jansen's book, and was Arnauld indeed holding to those same propositions? Finally, were those beliefs heretical? This commission lasted about a month. Arnauld tried to get a hearing, but they turned him away, and so the only thing he could do was to condemn the entire affair as illegal. He had his own supporters, however, and they strove to argue his case. But the game was rigged. Everyone knew what the outcome would be before it started, and on January 14, 1656, the faculty voted 130 to 71 to declare that Arnauld's *Seconde lettre* was indeed heretical. Arnauld's teeth had been pulled, at least for the time, and the best he could do was to make a public statement that they had misused his terminology, for his ideas could be found in the fathers of the church, from St. John Chrysostom to St. Augustine.

From this point on, Arnauld could not walk around freely for fear of being snatched by Mazarin's police. He left Paris quietly and moved to Port-Royal des Champs, out of the spotlight for a time. At this point, one of those great moments of history occurred: Arnauld, the dour theologian of Augustinianism, traveled back and forth to Paris not in his familiar black soutane, but in ruffles and laces, all fashionable, with silks and a wig, with shirts sporting "cuffs and tassels."

At Port-Royal des Champs, the faithful gentlemen gathered daily to discuss their options, which seemed to be thinning by the day. In the middle of January, Blaise Pascal, on his second retreat, joined them. They were not a happy gathering, for they were stumped. Arnauld tried to write another defense of their position, learned and lawyerlike, but none of the other gentlemen thought it was worth doing one more time. "I see that you find this text inadequate," Arnauld said, "and I agree with you." Then he turned to Blaise and said, "But you are young. You ought to do something."[66] Blaise agreed. Several days later, the group gathered again to listen to the opening lines of Blaise's *Provincial Letters*.

Sir:

We have been duped. Only yesterday, my eyes were opened; I had thought until then that the subject under dispute at the Sorbonne really mattered, and would have extreme consequences for our religion. All those assemblies of such a famous group as the theology faculty of the University of Paris, all those meetings which have given rise to such extraordinary and unparalleled events, give you an elevated impression of the proceedings, such that one can only believe that these men are dealing with an extraordinary subject.[67]

Now the theological argument had moved on to new ground, one that would later be trod by Voltaire. The debate was no longer a matter of careful theological reasoning, but of mockery. The modern age had truly begun.

The Jesuit Menace, Part 2

*As your scurrilities are daily increasing, and as you are
employing them in the merciless abuse of all pious persons
opposed to your errors, I feel obliged for their sake and that of
the Church, to bring out that grand secret of your policy, which
I promised to disclose some time ago, in order that you may all
know, through means of your own maxims, what degree of
credit is due to your calumnious accusations.*

—BLAISE PASCAL, *Provincial Letters*

Blaise Pascal single-handedly invented the myth of the crafty
Jesuit, a myth that has persisted to this day. He invented the
word *Jesuitical*, meaning "crafty," "sly," "devious," and one could trace
Pascal's hatred of the Jesuits on through Voltaire, up to the day when
Thomas Jefferson fretted with John Adams in his old age as John Carroll
imported a number of Jesuits to begin colleges across the country.
Jefferson said that with the coming of the Jesuits, the United States would
be beset by priests wearing disguises that "only the King of the Gypsies
could assume."

Not that the Jesuits didn't have their bad eggs. Indeed, they did, but
far fewer than the myth would indicate. In truth, what Pascal did was
to take an already existing prejudice against the Jesuits and give it a new

spin. The Jesuits were unlike most of the existing old religious orders, for they had a special relationship with the pope through their fourth vow of obedience, and this set the teeth of French nationalists on edge. The Jesuits were "ultramontanists," those who looked over the mountains to Rome for their instructions. Nationalism is a powerful force, and the French saw themselves as holding a unique position within the history of Catholicism. They had special privileges and special freedoms, and any attempt by the pope or his agents to undercut those privileges seemed to be an attack on the nation itself. Moreover, the Jesuits didn't act like most religious orders, for they were humanists through and through, children of the Renaissance and not of the Middle Ages. Ironically, had he known what the hubbub was about and not dealt with the Jesuits out of the myth, Jefferson would have agreed with them far more than he would have agreed with Pascal.

There are two basic impulses in the history of religious life in the Catholic Church. The first is the most ancient and sees the world as a shipwreck, as a place of sin and evil, so that the pious Christian would best serve the faith by withdrawing from the world to a place of safety, like the deep deserts of Egypt or monasteries like Port-Royal des Champs. The other impulse is based on the belief that the world is God's creation and that it is the Christian's duty to help in the conversion of that world. This was the view first of the Benedictines, then later of the Franciscans and Dominicans, and finally of the Jesuits. Didn't Jesus himself send his apostles out among the nations to convert the whole world? For this kind of religious, the act of hiding in the desert, working on your own spiritual perfection, seems like the act of the man who buried his talent in the ground because he was afraid of his strict master. The Jesuits believed that one could "find God in all things" and that the world was not a place to be abandoned in some introverted drive for self-perfection, but a place to be acted upon, a place to change the hearts and minds of the wavering masses through action.

In this sense, the Jesuits and the Jansenists were speaking two different languages. The Jesuits were born Molinists, not just because Molina was one of their own people, but because the exercises of St. Ignatius had set them up to be such. In the exercises, there is a meditation on the Two Standards, where the retreatant arrives at a decision—whether to stand under the banner of Christ or under the banner of the devil. This decision was always a freely chosen one, one that assumed that a person had free will, and that that free will was radical. The individual soul was not a puppet dangling on a string held by her Almighty Creator, but was a free person making a free choice. Indeed, the entire spiritual exercises are designed to lead a person to that moment of decision. What could be more different from Jansenist Augustinianism? If there was a beef between the Jesuits and the Jansenists, it was because the Jesuits were not Augustinians and because, though they admired Augustine from a distance, they did not accept his evaluation of human nature any more than they accepted the same evaluation from Luther and Calvin.

Pascal vigorously attacked the "casuistry" of the Jesuits, by which he meant the application of general moral principles to individual cases. Most of these cases were hypothetical, and books on casuistry, which were common in the seventeenth century, were simply instructors' manuals for confessors, Zagat's guides to the restaurants of sin.

What Pascal objected to was an attitude that he called "laxity," whereby Jesuit confessors would use reason to excuse the worst sins. Often the Jesuits did this as a tactic to demonstrate to the sinner that God's love was in their lives, that they had "sufficient" grace for penitence, and that God would supply them with the grace they needed for conversion. Jesuits were not above scaring the wits out of sinners, as the tradition of Jesuit theater in Rome suggests, but they did believe that God was on the side of everyone, even the worst sinner, and not just on the side of the elect.

Pascal and his crowd saw the sinners as already abandoned, unless of course some wondrous infusion of efficacious grace popped into the sinner's life, as it had in the life of the apostle Paul and even in the life of Augustine. For a Jansenist, sin was everywhere, and the life of penance required that one seek it out, not excuse it, for even the best of us live our lives in the red, in debt for the sin of Adam. The whole idea of taking a universal moral principle and applying it creatively to individual cases was suspect in its creativity.

There are eighteen letters in Pascal's *Lettres provinciales,* written under
the pseudonym Louis de Montalte. The volume is a masterpiece of mockery and was understandably among Voltaire's favorite bedtime books. He
always carried a copy with him wherever he went, even though he couldn't
stand Pascal in any other costume. The first few of these letters take on
the persona of a confused man about town, writing to his friend in the
provinces about the odd theological goings-on in the capital. There are no
villains, just silly people, holding silly ideas. Naturally, the Jansenists come
off as the only rational bunch in the barrel. Pascal makes great sport of intellectuals, especially the doctors of the great Sorbonne who were about
to evict his friend Arnauld. In his first letter, Pascal puts words into the
mouth of an unnamed theologian: "'Hold there!' said he. 'One must be a
theologian to understand this question. The difference between the two
parties [the Sorbonne-istas and the Jansenists] is so subtle, so excruciatingly complex that we can barely tell the difference ourselves—you will
find it too much for your limited powers.'"68

These excruciatingly complex and subtle differences, Pascal argues, are
creating a war within the church. "What picture can I present you of
the Church batted back and forth between these conflicting sentiments?"
Pascal said in the second letter, written on January 29, 1656. "Just like
the man who, leaving his native country on a journey, is encountered by
robbers, who inflict many wounds on him and leave him for dead. He
sends for three physicians who live in the neighboring towns. The first, on
probing his wounds, announces them mortal and assures him that none
but God can restore his life to him. The second, coming after the other,
chooses to flatter the man, tells him that he still has sufficient strength to
reach his home and, abusing the first physician who opposed his advice,
decides to ruin him."69 The third physician, after being entreated by the
wounded traveler, merely sides with the second and attacks the first. The
barely disguised allusion to the Parable of the Good Samaritan is classic
Pascal, as is his oversimplification of the problem.

By letter three, published on February 9, 1656, Pascal is still trying to
"understand the whole affair in a pleasant way" in the beginning, but
by the end of that letter, his fangs come out. By that time, the pressure

on the Jansenists was becoming a crisis, and the Jansenists were running scared. From that point on, the rest of the *Provincial Letters* becomes a direct assault on the casuistry of the Jesuits, often taking examples, out of context, from manuals written by Spanish Jesuits.

"Casuistry" simply means making moral judgments from individual cases, and it has its roots in Jewish moral thought. In Catholicism, when the sacrament of Penance moved from the *forum externum*—that is, from a public judgment of actions and a public assignation of penances—to the *forum internum*—that is, private confession—the confessor was no longer merely the judge and jury, but also the spiritual adviser, whose job it was to guide the penitent back to the fold. It was at this point that the professional spiritual counselor was born, a job that has come down to us in its medical form as the psychiatrist and psychologist. One of the greater tasks of such counselors was to fight scrupulosity, the unreasoning sense of one's inherent evil and the unreasoning fear of damnation. In doing so, the spiritual father, or adviser, became the understanding guide, the one who took intent into account, who defined sin along narrow lines, who tried to show the penitent the place where God lived in their hearts.

By the sixteenth century, casuistry had become a recognized science, one that was nearly destroyed by the controversies with the Jansenists. The heart of this controversy was over the doctrine of probabilism, which held that when there is question solely of the lawfulness or unlawfulness of an action, the confessor can opt to follow a probable opinion in favor of liberty, even when the other opinion might be more likely. In other words, the Catholic Church rigged the game and leaned in the direction of moral liberty, which drove the Jansenists mad. This tendency—among Jesuit confessors, especially—they called "laxity." Jansenists, therefore, were "rigorists," whereas the Jesuits were "laxists." Needless to say, eventually the pope condemned extremes on both sides.

Pascal was one of the worst practitioners of rigorism, even to his death. People had to behave as close to perfectly as possible. Moreover, his attacks on the Jesuits in the remaining fifteen letters became ever more vicious, and ever less honest in his association of all casuistry, even Jesuit casuistry, with the worst abuses of the science. Certainly there were abuses, but those abuses were not typical of the practice. But the success of the *Provincial Letters,* success that derived mainly from its mockery, created the myth of the clever Jesuit and associated all casuistry with a cynical practice of making excuses for sinners. Jesuits were even willing to wave off homicide, Pascal argued. They were willing to do this in order to purchase political might, to buy their way into the halls of the powerful. "Because of the fame they have acquired in the world," Pascal wrote in the fifteenth letter, written on November 25, 1656, "they can defame people without degrading the justice of mortals; and, on the strength of their self-assumed authority in matters of conscience, they have assembled maxims for themselves that enable them to act without any fear of the justice of God. This, fathers, is the fertile source of your base slanders."[70]

The last letter appeared on March 24, 1657. Only a year later, the true identity of the author came out, and François Annat, the Jesuit spiritual adviser to the king, declared that Pascal was a heretic. By this time, however, Pascal's health was fading, and he had become a frail shadow of the man who had so ferociously written the *Letter to a Friend in the Provinces.* Nevertheless, the letters had already become the literary hit of the decade and the new voice of the Paris underground press, the same press that would feed the fires of revolution in a later age. Louis XIV banned the *Provincial Letters* in 1660, and it was put on the list of forbidden books by the Inquisition. The king quickly ordered all copies gathered and burned.

But the *Provincial Letters* did survive and became a rallying point for a later generation of *libertins érudits,* the French Deists of the next century. In spite of all his good intentions, Blaise Pascal the über-Catholic had handed a terrible weapon to the enemies of the church: the grinning face of mockery, to be used not against the spoiler Jesuits, but against the faith itself.

The Miracle of the Holy Thorn

The Age of Miracles is forever here!

—THOMAS CARLYLE

To me every hour of the light and dark is a miracle,
Every cubic inch of space is a miracle.

—WALT WHITMAN

Men reject their prophets and slay them, but they love their
martyrs and honor those whom they have slain.

— FYODOR MIKHAYLOVICH DOSTOYEVSKY

T he rapid publication of the *Provincial Letters* on January 27, February 5, and February 12, 1656, was an underground sensation. The letters were a hit among the educated middle class, and the fact that they sent Cardinal Mazarin's secret police scurrying around looking for the author, or at least the printers who had printed them, did not harm their popularity in the slightest. The Fronde was still bubbling along, and the queen and the cardinal were far from popular. The police found the printers from time to time, wrecked the print shops, and harassed the printers, so that they had to move their operations to more secret locations. They did not find out the true identity of the author of the letters until much later, though the Jesuits had their suspicions.

Pressure on Port-Royal increased so that in mid-March the crown issued an edict ordering the solitaries to leave Port-Royal des Champs at once. According to Mère Angélique, the valley had become a vale of tears. The gentlemen, some of whom had lived there for twenty years, and the fifteen young students in the *petites écoles* left the old monastery with great sadness. Suddenly, two miracles occurred, one worked by Mère Angélique and the other, according to the Jansenists, worked by God. Mère Angélique's miracle was far more mundane, though quite miraculous in its own right. On March 20, her agents within the royal palace informed her that the nuns would soon be expelled from Port-Royal, not only the facility in the country, but the one in Paris as well. Somehow, her agents had seen a copy of the document sitting on the Queen Mother's dressing table. What else could she do? Mère Angélique's political resources had been stretched to the limit, and the forces arrayed against Saint-Cyran's spiritual children were growing more powerful each day. And it wasn't just the Jesuits. In fact, it wasn't mainly the Jesuits, though one would think that from the *Provincial Letters*. By that point, the Jansenists could count among their enemies the young king, the Queen Mother, Cardinal Mazarin, the majority of the diocesan clergy in France, Vincent de Paul, Jean-Jacques Olier, and finally the pope. So Mère Angélique took her woes to God, and for the next three days and three nights she prayed before the Blessed Sacrament in the chapel of Port-Royal des Champs.

Meanwhile, back in Paris, the sisters prayed their own prayers to save their way of life. Staying at Port-Royal de Paris at the time was Gilberte Perier's ten-year-old daughter, Marguerite, Blaise's favorite niece and godchild. She was a young student, or *pensionnaire,* at the school there who had come to Paris from Clermont with her mother on the outside chance of finding a cure for an eye disease that tormented her. Her left eye had become infected, and a fistula, filled with pus and blood, swelled underneath it. Worse yet, the eye stank with infection, and no doctor in Clermont could do anything about it. Gradually, the infection spread throughout her body, and she spiked fevers. Both Blaise and Jacqueline did their best to care for the girl, and sent her to the best physicians they could find, but none of them had an easy cure either. The doctors were certain that the infection would spread to other parts of the little girl's body, especially her nose, and they suggested ever more radical treatments. Finally, they recommended an operation on the wound, essentially by taking a hot poker and cauterizing the fistula. They could cure the wound by doing this, they said, but it was also possible that they would kill the girl. Blaise sent word to his brother-in-law about what the doctors had said, but the thoughts of hot pokers jabbing the eye of his little girl galled poor Florin, and he wrote back asking that they postpone the operation until he had the chance to come to Paris and make a decision for himself.

It was March 24, and Mère Angélique was just finishing her vigil before the Blessed Sacrament, praying for a sign of God's favor for her beleaguered convent. As a pious act, Jacqueline Pascal wrote, the sisters at Port-Royal in Paris had placed on their altar a "very beautiful relic in which was encased in a little sun of silvergilt a splinter of a thorn of the Holy Crown." According to legend, this sliver of thorn was a tiny piece of the crown of thorns placed upon the head of Jesus by the Roman soldiers. Obviously identifying with the passion of Christ, the sisters of Port-Royal had spent the day venerating the relic. They brought the students along with them to join them in their prayers. As little Marguerite approached the holy relic to kiss it, on an impulse, the sister in charge of the students, Sœur Flavia, took the relic from the altar and touched it to Marguerite's eye. That evening, Marguerite came to her and showed her that the fistula was gone. "My eye is healed, Sister," she said. Sœur Flavia ran at once to Mère Agnès, who was acting superior at the convent while her sister Mère Angélique was in the country. Mère Agnès ordered that everyone who knew about the incident keep quiet until it could be properly investigated.

It wasn't until March 29, the following Wednesday, that the sisters told Marguerite's uncle Blaise, and asked him to bring the physician along with him. Both showed up on March 31, and once the physician examined Marguerite's eye, he announced that the eye was healed and that Marguerite had completely recovered. He had no idea how this had happened but wanted to see Marguerite again soon, after he had given himself a chance to think things over. Meanwhile, Jacqueline summoned Marguerite's father from Clermont. On April 7, in the presence of the sisters and of little Marguerite's father and uncle, the surgeon declared that the girl's eye had been cured by an act of God.

But the opinion of one physician, though confirming, was not enough. Everyone who knew about the miracle understood that miracles were complicated affairs. On one hand, the healing of Marguerite's eye may well have been the sign of God's favor that Mère Angélique was praying for; on the other hand, if Port-Royal proclaimed the miracle too widely, they might be accused of creating a hoax. In seventeenth-century France, miracles were highly political, and everyone involved knew that. For that reason, Florin Perier and Blaise Pascal both decided to keep the event quiet until they could gather more medical support. They invited seven doctors, the ones who had treated Marguerite's eye during the previous months, to gather at Port-Royal on April 14, which happened to be the Wednesday of Holy Week. Each doctor examined her eye, and after a long discussion among themselves, all seven of them signed a document proclaiming the miracle: "Since this sort of cure of so severe a disease effected in an instant can only be termed extraordinary, to the degree that it can be understood we think that it surpasses the ordinary forces of nature and that it could not have been done without a miracle, which we believe to have been genuine."[71]

As happy as the people at Port-Royal were about the healing throughout the Easter season, the Pascals and the Periers foremost among them, they decided to keep the event quiet until they could figure out whether it would help them or hurt them. However, no one told the doctors, and they spread the word throughout Paris, so that within a very short time the entire city was electric with talk of the Miracle of the Holy Thorn. Even people at court, those who would once have counted themselves enemies of the Jansenists, were fascinated by the stories. Queen Anne sent the king's doctor to examine Marguerite, and he seemed to concur with his colleagues. Things began to turn around for the sisters at Port-Royal, for even as some of the most radical enemies of the Jansenists—namely, the priests of Saint-Sulpice—complained that the miracle was a hoax, the king's physician simply told them that it seemed like a miracle to him, that "the little Perier" appeared to have been healed. Soon after, on May 27, a formal inquiry was launched by the archbishop of Paris. The vicar general interviewed Marguerite, Florin, Blaise, the physicians, and a number of the sisters who had witnessed the alleged miracle. On October 22, the committee, composed of five theologians, declared that the healing of Marguerite's eye was indeed a miracle, an act of God, and a supernatural intervention into the world. At Port-Royal, Père Singlin presided over a solemn Mass of thanksgiving, where both the relic of the holy thorn and little Marguerite Perier were placed on display. What the ten-year-old girl may have thought of her sudden burst of fame is uncertain.

Pascal's Wager

If you ain't just a little scared when you enter a casino, you are either very rich or you haven't studied the games enough.

—VP Pappy

Gambling: The sure way of getting nothing from something.

—Wilson Mizner

The subject of gambling is all encompassing. It combines man's natural play instinct with his desire to know about his fate and his future.

—Franz Rosenthal, *Gambling in Islam* (1975)

By the spring of 1658, the attacks on Jansenism had quieted for a time, the result of the Miracle of the Thorn. "When there are parties in dispute within the same church," Pascal wrote, "miracles are decisive."[72] Gilberte later wrote that after her daughter was so mysteriously healed, her brother became fascinated by miracles and visited every church in Paris that sported a relic with some miracle attached. He went to every liturgical festival, every High Mass he could find, and even kept a record of the church celebrations in the city so he could attend as many as possible.

Though he never quite considered himself a member of Port-Royal, he was still a partisan, was deeply involved with the movement, and was in total agreement with its leaders. Indeed, he even shared some of their danger. Had Cardinal Mazarin's secret police ferreted out the true identity of the author of the *Provincial Letters* sooner, Pascal could have found himself wasting away his last days, somewhat shortened by the cuisine, in the Bastille. But because Mère Angélique and Père Singlin still suspected him of worldliness, he was never fully embraced by the movement, and so Pascal began to define his life in terms of his own commitment to the faith, by which he meant the faith as read by Augustine and Cornelis Jansen, regardless of what the purists of Port-Royal thought of him. In the period leading up to April 1658, during his last year of good health, he delivered a series of seminars to the gentlemen at Port-Royal des Champs. Here, he outlined his plan to write a comprehensive apology for the Christian religion and told them that the old proofs for the existence of God were abstractions beyond the ken of the average person. Too many philosophers—and Pascal must have had Descartes in mind— gloried in abstractions, while such abstractions were of little use for the common man, and would have little impact on their lives. For Pascal, the ontological proofs for God's existence were so far from human life that they meant little.

A true apology for the faith, Pascal argued, would have to take full account of the human condition, which is that human beings are both great and wretched at the same time, that we are great because we have reason and thus can see the world for what it is, while we are wretched because we have no real power to change it or ourselves. His apology was, in essence, a theology of moral powerlessness. We are all sinful, base creatures who desire heaven most of all. This is our glory and our most wretched suffering. Such an apology would lead the questing mind from a realization of the truth of the human condition, with a full acknowledgment of all that is dark and ugly about humanity, into a new light based on penance, a penance that could introduce the soul to a cure for this helplessness, and from there onto a path for obtaining that cure. For Pascal, any argument for the faith must appeal to the "three orders" of the

human condition—the intellectual order, the moral order, and the physical order. One cannot argue the existence of God without also appealing
to the heart. The word *heart* here means more than just the emotions. It
means that deep abiding sense of wonder and fear that constitutes human
consciousness as it confronts its own insignificance in the face of God and
of the universe.

"When I think about the shortness of my life," Pascal said, "melted
into the eternity that came before me, and into the eternity that will
come after . . . and the insignificance of the space I fill and even see, I'm
lost in the infinite vastness of that space that lies beyond, that space of
which I am ignorant and which has no knowledge or care of me. I'm
frightened and astonished to awaken in this place rather than that, and
I see no reason why I should be here and not there, now and not then.
Who put me here? By whose order and direction have this place and time
come to me?"[73]

For Pascal, this was the greatest moment of discovery of all, a far greater
moment than the discovery of the vacuum or of the laws of probability.
For many contemporary people, this same insight leads to atheism, for
how could there be a God who loves us in the face of our insignificance?
For Pascal, it led to a deep, abiding faith in a God who had disappeared
from the earth and had chosen to remain hidden. But this *Deus absconditus,* this hidden God, still abides in us as the great desire of our hearts, for,
as St. Augustine said, "My heart will not rest until it rests in you." Outside
of that God, for Pascal, there was no meaning in life, no purpose, and no
point to any study, or any science.

It was Pascal's desire, therefore, to create an apology not of the mind
but of a heart. In this, he was the great precursor to existentialism, which
tried to gaze at the human condition with an unflinching heart. But in its
atheism, or at least the atheism of its most famous practitioners, existentialism merely encouraged people to bravely face the meaninglessness of
their lives, the meaninglessness that Pascal abjured, and to courageously
attempt to create meaning out of nothing. For Pascal, such atheism was
impossible, and worse, it was a fool's errand. Meaning cannot be created
from emptiness, nor can it be found by the moral courage required to

face nothingness. To look into the human heart and the human condition for him was to ultimately find God.

Because he based his apology on the complete grand opera of the human condition, he wove into it those very diversions that he had said kept people from examining their lives and therefore kept them from God. The chief of these was gambling, something he knew a great deal about. How many hours did he spend watching the nobility of France gambling away their fortunes, and daintily laughing in the face of their own folly? Certainly, he was an astute enough observer to see the religious power embedded in the act of rolling the dice, as the knights and lords and ladies risked everything they valued in *le hasard,* tilting with the bitch goddess Fortune. If they could risk their wealth, their honor, and their reputations as part of the game, a mere sport, would they not also be willing to risk the same on the possibility of eternal life? Here then was Pascal's practical argument for the existence of God. In it, he showed that religious belief was not irrational, as the Pyrrhonists, the skeptics, and the atheists would say. Indeed, to believe in God was rational both in the mind and in the heart. This, then, is the argument of his famous wager.[74] The argument is a dialogue addressed to the rational doubter, to the man of the world, a man very much like Pascal's own father. It begins not with faith or with doubt, but with a simple assertion that God either exists or doesn't. If God did exist, then this God would be unfathomable, infinitely unknowable, beyond our imagination. And therefore, Christians should not be accused of stupidity if they can't give a rational account of their faith. Would one expect a God who was so far above us to be accountable to reason?

Okay, say the nonbelievers, so believers might be excused for their irrationality. But should not an honest, rational doubter be accused of irrationality if he bought into religion?

Pascal responds that reason is powerless to decide the existence or nonexistence of God. In doing this, he tacitly rejects all the cosmological proofs of Aquinas and the ontological proofs for God's existence put forth by Anselm and Descartes. The best way to understand the issue is to treat it like a game:

Let's examine that point, then: let's say that God does or does not exist.
Which side should we choose? Reason is powerless before such an issue.
There is an infinite abyss separating us. At the far end of this infinite uni-
verse, a coin is tossed—which will turn up, heads or tails? What will you
wager? Relying merely on reason, you can't decide. You can't rationally
bet either way, for you can't defend either choice.

Thus, don't call people who have made a choice fools, for you know
nothing about it.[75]

But then the doubter says that people can be blamed for making any
choice at all, that it would be best not to choose. Pick either side and you
are a fool.

This is where Pascal steps into a new territory, and sets the stage for existentialism. His argument shifts its focus from the order of the world to
the common lot of humanity. He says to the doubter that it's the human
condition to choose, that we can do nothing else:

But, I say it's necessary to bet. You cannot avoid it, for you are already
launched on the waters. This being the case, which one will you take? How
will you decide? Come now, since you must choose, let's consider which one
is of less importance to you. You have two things to lose—the true and the
good, and two things to stake—your reason and your will, your knowledge
and your bliss, and your nature has two things to shun—error and misery.
Since you absolutely must choose [by living, you cannot avoid it], your intel-
ligence will not be offended by one choice any more than by the other. That's
one point settled.

So we are in the game; so we have to set the outcomes: if you win, you
win everything in this game, and if you lose, you lose nothing, so why
not play? So how do you set your bet? That depends on what's at stake.
In this game, says Pascal, the prize is eternal life. By belief in God, what
we are risking is a lifetime of meaningless pleasures. We are accepting a
life of spiritual discipline in the hope of gaining an infinite reward. One
would be irrational to refuse such a bet. If God exists, and you believe in

God, then throw the dice and you win an eternity of life, love, and joy. If God doesn't exist, and you believe, and you throw the dice, then all you lose is the pile of meaningless pleasures you were sitting on. If God does exist and you don't believe, then you give up all hope of eternal life, and worse than that, you will one day find yourself in hell. Therefore, it is a prudent and reasonable thing to believe in God, because you could gain everything and risk very little. There isn't a casino in Vegas that could give such odds.

The doubter says that even if he wanted to believe, his own nature wouldn't allow it. He can't just force himself to believe if he doesn't. Pascal responds that action comes first. Act as if you believe, and belief will follow:

> *It's true, but at least wake up to the fact that your inability to believe comes from your passions, since reason induces you to believe but you still can't. So, don't strive to persuade yourself by counting up proofs of God's existence; strive to diminish your passions. You want to find faith, but you don't know the way. You want to cure yourself of your unbelief and you're asking for remedies; learn from those who were once tied up like you and are now throwing the dice. They are people who know the path you'd like to follow; they are people cured of a disease from which you'd like to be cured—follow the way they started on.*
>
> *They acted as if they believed—they took holy water, they had Masses said, and the like. That will make you believe quite naturally, and will make you more pliable to the faith.*
>
> *"But that's what I fear." Why? What do you have to lose?*
>
> *How will you be harmed by choosing this path? You will be faithful, honest, humble, and grateful; you will be full of good works, and will become a true, good friend to those who know you. What will you lose? Noxious pleasures, vainglory, and riotous times, but these loses will be easily supplanted by other, greater joys.*

Of course there are holes in what Pascal says, but most of those come from misunderstanding his project. Those who want to argue against it as they would any philosophical idea are playing Descartes' game and not Pascal's. They are back to metaphysics, looking for an argument that is so compelling that the bystanders must believe its conclusions or accuse themselves of being irrational. For Pascal, such arguments, though they look nice, lead us into orbit and leave us spinning there, without rooting us to the earth of human life. To understand his argument, you must think not like a philosopher but like a gambler. What Pascal is doing is applying to questions of the existence of God the old rule of expectations that he suggested to Fermat in his letters on the gambler's ruin. Does he claim any absolute proof for God's existence? Quite the contrary: he acknowledges that such proofs are illusions. What he tried to do, within the context of seventeenth-century France and the religious climate of that time, was to demonstrate that if you did believe in the Christian God, especially the God of St. Augustine, then you weren't an idiot, that you were taking a calculated risk that had a good chance of succeeding. This argument can be found in other religious contexts, most notably Hinduism, and can really be judged only within the context of each religious culture.

In Pascal's argument, there are four possibilities:

1. You believe in God, and God exists. This would mean you could go to heaven, and your winnings would be infinite and everlasting.

2. You believe in God, and God doesn't exist. In this case, you die like everyone else and rot in the grave like everyone else, and what you lose is some wild times on earth, which compared to eternity is nothing, and your loss is negligible.

3. You don't believe in God, and God doesn't exist, in which case, see possibility 2.

4. You don't believe in God, and God does exist, in which case, you are in a lot of trouble. You go to hell, and your loss is infinite.

Given the outcomes and the odds, therefore, it would behoove a betting man to bet on God. Pascal here is applying game theory to theology. Certainly unconventional, certainly puckish, but puckish with a dash of genius. Once you accept the rules of the game and the context of the game, you'd be the worst kind of donkey not to believe in God, Pascal's God. This is because the first possibility, believing in God, dominates the last, not believing in God.

Now, philosophers, being that sort, have been merrily punching holes in Pascal's argument for centuries. Almost all of their jabs turn on some act of stepping out of the context of Pascal's game—that is, the context of a Christian society that was losing its faith—and criticizing the whole thing from that outside position, like the uninitiated kibitzer who comes to a bridge tournament and shouts, "Well, that's stupid!" every time somebody puts down a card.

First of all, they argue, what if God isn't into rewarding or punishing? What if God doesn't do things like that? Pascal's argument would then fall apart. After all, they say, doesn't Pascal assume a Christian God? What if God is a Hindu God, or Muslim, or even maybe Zeus? Couldn't Pascal's wager also be used to encourage belief in Chemosh? Or Odin? Or New Age pantheism? All of these criticisms are fine, if you want to play another game, some game other than Pascal's. It is easy enough to stand outside of his argument and propose other possibilities, but to do so would be to miss the point. Pascal made no claim to metaphysical or even mathematical compulsion. He made the claim only that it was prudent to believe in the Christian God, at least from inside the world that he lived in. His argument was not made to logic-chopping philosophers, but to *honnêtes hommes*, gentlemen gamblers running the odds at Vegas.

One thing that Pascal does assume is that there is at least a possibility that God exists. Therefore, his argument would have little effect on strong atheists, because they argue that there is no possibility that God exists. But of course, this is as much a belief statement as is Pascal's Jansenism, and can be relativized in exactly the same way. People can stand outside of the atheist's game and shout, "That's pretty stupid!" just as they can outside the Christian game.

Then, there was the liberal Protestant criticism of William James, who didn't much care for Pascal's Catholic hocus-pocus.[76] Pascal would have you believing in God out of a hope for a reward and fear of punishment, says James, and this, says James, isn't real faith. A true, pure faith should have none of that, and should be held freely, and without strings attached, out of a desire to do good. Thank you, but Pascal is not talking to strong believers, but to wavering believers, and most especially to cultural Catholics, to those people who had been brought up in the faith but could not get beyond their own immediate desires. Pascal's wager is therefore a prudential one, aimed at the tepid and the lukewarm, those who would go to church if they could see a percentage in it. The kind of purity that James is demanding is something that Pascal would address somewhere down the line. His wager is not evangelization, but pre-evangelization, and James's argument smacks of the fellow who went to the Happy Kangaroo Play School and demanded to know why the children weren't studying ablative absolutes.

But now comes the real test, the only true criticism of Pascal's wager. It is a Jansenist criticism that, following Augustine, would question whether people are free to choose the good in any way, without efficacious grace. So why talk to these waverers at all? Why write an apology for Christianity if the elect have already been chosen and will believe out of efficacious grace, whether they choose to or not? Would this apology not be a fool's errand—to try to convince such lukewarm folks when efficacious grace is so obviously absent? Moreover, wouldn't there be a possibility that the apologist was going against the will of God in doing so? There is something rather Jesuitical in the entire project, something taken from his great enemies, which had entered and infected his Jansenist worldview. Maybe Mère Angélique and Père Singlin were right after all. Buried in his argument is the notion that people can decide on their own beliefs, that they have the power to choose the good and to change their lives. Perhaps at bottom Pascal was a humanist after all.

Port-Royal Agonistes

Grace me no grace, nor uncle me no uncle;
I am no traitor's uncle, and that word "grace"
In an ungracious mouth is but profane.

—SHAKESPEARE, *Richard II*

Grace! 'tis a charming Sound,
Harmonious to my Ear!
Heav'n with the Echo shall resound,
And all the Earth shall hear.

—PHILIP DODDRIDGE (1702–1751), *Hymns Founded on Various*
Texts in the Holy Scriptures (1755)

Grace under pressure.

—ERNEST HEMINGWAY, *New Yorker* (NOVEMBER 30, 1929)

Blaise spent the summer of 1660 staying with Gilberte and her family in Clermont. While there, he received a letter from Pierre Fermat inviting him to Toulouse for a visit, but Pascal declined, saying that his health was not strong enough to permit it. Moreover, he said, he no longer considered geometry to be anything more than a craft—a noble craft, to be sure, but a craft nonetheless. His religion had finally trumped his science. Thus he and Fermat, who had been collaborators through so many years, on so many projects, never met in the flesh. In the end, Pascal regretted it, not because Fermat was a man of mathematics, but because he was a man of honor and integrity—a good friend, even from a distance. Though his life as a scientist and mathematician was far from over, the young ambitious careerist who had so ardently defended the existence of the vacuum had become something else entirely. His values had changed, and he was a new man.

Instead of traveling to Toulouse, Pascal left for Bourbon to take the curative waters, and from there went on to Poitou to spend Christmas with the duc de Roannez and his family. On December 13, 1660, while Blaise was still in Poitou, the young king called the officers of the Assembly of the Clergy to him at the Louvre and announced to them that he had determined to bring about the end of Jansenism. Cardinal Mazarin stood nearby, but it was obvious to everyone that he was sick unto death, and that this decision was not his doing.

It is likely that news of the king's announcement hit Poitou fairly quickly, putting a chill into Pascal's already wasted body. By the time he returned to Paris, the city was buzzing with news of the death of the cardinal, and of Louis's newfound determination. For the Jansenists, this was like the tolling of a bell, not only for the cardinal but for themselves, for he was the one person who had been holding the young king's opposition in check, and although he was no great friend of Port-Royal, he did not wish to go to war with them, either. But all of that ended as the cardinal lay dying in Vincennes, when Louis affirmed the rumor that he would not replace Mazarin but would take on the duties of first minister himself. In doing so, he was declaring to the nation and to the world that he and only he would be the master of France.

The partisans of Port-Royal shuddered at the announcement, for the young king was no friend of theirs. Back in 1657, Louis had taken counsel from his pious mother, who was then under the influence of Vincent de Paul, and announced his opposition to Port-Royal and everything it stood for. Now that he had come into his own power, he had declared them to be his enemy. In France, there would be one king, one faith, one law, and that would be Louis XIV on all three counts.

By that February, the Assembly of the Clergy published their document, including a formulary one that had the legal power of an oath, and required that all priests and religious sign it as a proof of their orthodoxy. This formulary contained nothing less than a complete repudiation of Jansenism, so that when the Port-Royalists signed it, they were signing their lives away. By the king's command, the peace brought on by the Miracle of the Thorn was over. On April 23, an officer of the court appeared at Port-Royal in Paris, handed Mère Agnès Arnauld an official document, and then announced to her that the monastery could no longer accept postulants until further notice, that they were required to close the convent schools, the one in Paris and the one at Port-Royal des Champs, and that they were to send the nearly seventy students, or *pensionnaires,* back to their homes. Of course, this included the two Perier girls, Blaise's nieces, as well. Next, the king sent another court official to the country to close Lemaître's *petites écoles.*

Along with this, the crown removed Père Singlin as the spiritual director for Port-Royal. Mère Angélique, who had often taken to her bed, stricken, when opposed by a stronger personality than her own, was suddenly stricken once again at the prospect of losing all she had created, and this time in earnest. She died in late summer, after calling her sisters to her and consoling them about their dark future. Then, just before she died, she said that she expected it would be a long, terrible eternity.

As Mère Angélique lay ill, her brother Antoine and the other Jansenist leaders gathered to discuss their strategy. What they wanted was a new document, drafted by their friends and supporters among the diocesan clergy, that they could sign and that would yet maintain the distinction between fact and law. Pascal was involved in this conference because he had written the *Provincial Letters,* and had therefore had taken several good swipes at the Jesuits and their supporters. In their document, they looked for some kind of accommodation that would allow them to sign the formulary and yet maintain their beliefs. Luckily, the original formulary had gaping holes left in it, so that all of Port-Royal would be able to squirm through.

The document was ready by the end of May, and it called for a "complete and sincere respect" for the church's teaching authority, even in its condemnation of the five propositions. It even forbade anyone connected to them from "preaching, writing, or disputing anything in a sense contrary to the assent of faith." However, Antoine Arnauld, Nicole, and the others made certain that the document did not include an admission that the five propositions were indeed found in the *Augustinus,* which allowed them to preserve their carefully constructed distinction between fact and law. Satisfied, Arnauld and Nicole then recommended to the sisters at Port-Royal de Paris that they sign the formulary. At first the sisters resisted, but then Père Singlin secretly consulted with them and convinced them. They signed on June 22.

But this wasn't enough for Louis, who was determined to bring the Jansenists to heel. He quickly made it clear through his royal council that this new document had too many loopholes, and he commanded the vicars general to go back and draft a new formulary that would keep the Jansenists from squirming out of the cage he had put them in. The king wrote a letter in early July to the Vatican, asking the pope for a new statement, one that would tighten the ambiguities of the first formulary. The pope, Alexander VII, responded quickly for a pope, as quickly as the middle of August. By October 31, 1661, bowing to pressure from the king, the vicars general in Paris issued a new formulary. They were not happy about it, because they believed that the pope had transgressed on their Gallican liberties, but they did it anyway. The rub came about halfway through the formulary:

> . . . *I condemn with my heart and my lips the doctrine of the Five Propositions of Cornelius Jansenius, as found in his book titled Augustinus, condemned by two popes and the gathering of bishops; and whose doctrine is not that of St. Augustine but only of Jansenius, who has badly misunderstood the true sense of this holy doctor.*

Thus, in one stroke the papacy condemned Augustine's teaching on original sin, correctly interpreted by Jansenius, while praising this "holy doctor" and claiming that he was badly misunderstood. This new formulary was published on November 20, and read out in all the churches in Paris. At the end, there was a little postscript saying that priests and religious had fifteen days to sign it. The "or else" was understood.

Pascal was furious, not because his Jansenist friends had instructed their followers to sign the formulary, which was bad enough, but because of the effect this would have upon the sisters, most notably his own, residing at Port-Royal. This was an injustice, he believed, a persecution, for by demanding that they sign the formulary repudiating the theology of Jansen, and therefore of the abbé de Saint-Cyran, the king, the pope, and everyone else concerned was asking them to give up their faith. These were pious women, not theologians, and they were new to subtle argumentation and dissertations that went nowhere. They had settled down in their convents to live lives of penance and holy piety and were no threat to anyone. They were more than ordinary women; they were *holy* women. And to force them to repudiate the theology of Cornelis Jansen under the threat of closing their convents, confiscating their dowries, and sending them home to families where they no longer belonged was a crime; more than a crime, it was a sin.

But there was more to Pascal's reaction than this. His concern for the sisters was real, but moreover, by asking the priests and the nuns of Port-Royal to repudiate their faith in Jansen's theology, the church was also asking Blaise Pascal to repudiate the night of fire, for it was from that experience that he had learned that the true God was not a God of the philosophers and theologians but of the prophets. The God he had encountered was the God of Scripture, of Jesus Christ. In this encounter, he had learned about the wretchedness and the nobility of the human race. Everything that he had come to believe, everything he had written about in his *Pensées*, preparing for his great argument against the atheists, was being attacked by the Jesuits and their surrogates. And then, disaster—the greatest disaster he had yet faced.

On October 4, 1661, Jacqueline Pascal, Sœur Jacqueline de Sainte-Euphémie, died; her brother believed that she had died of a broken heart. Though he too would be dead within a few months, his anger with his old friends at Port-Royal for their theological maneuvering, and for their slippery recommendation to the sisters that they should sign the document, had become white-hot. In 1660, when the first stage of the persecution had begun, Jacqueline had been sent to Port-Royal des Champs, to get her out of the line of fire, and to have her act as subprioress and novice mistress. When the second wave of persecution hit in June 1661, Jacqueline was one of the few who resisted signing the formulary. She believed it to be an injustice to ask the sisters to sign even the first of the formularies, the one that gave the Jansenist party so much wiggle room.

In her anger, she sent a letter to Antoine Arnauld protesting his decision. Little Jacqueline Pascal, Blaise Pascal's baby sister, was apparently ready to start the revolution. "I am prepared, with the help of God, to die confessing my faith during these present sufferings. What are we afraid of?" The closing of the convents, the loss of their dowries, the loss of Port-Royal itself—these were not important. What was important was that they remain faithful to the teachings they were given. "But we can rest secure within the simple boundaries of our sorrow and of the meekness with which we shall accept our persecution. The love with which we shall embrace our enemies will tie us unseen to the church."[77]

Eventually, Jacqueline signed the first formulary, but she never quite reconciled herself with it. Like her brother, she suffered from fragile health, and the stress of watching everything that she had loved collapse around her finally broke her spirit, and she took sick. Within a few months, she lay on her deathbed. After she died, Blaise railed against his old acquaintances at Port-Royal for their dithering. Would Jacqueline still be alive if they had been willing to fight on? Luckily, she had died before the second formulary could be presented to her, and so she was spared that torment.

Blaise's anger exploded when Arnauld and Nicole decided that they could sign the second formulary of October 31. After all, no one was attacking St. Augustine, they said, or his doctrine of efficacious grace, so even if the church had misunderstood the *Augustinus,* none of the church authorities had asked them to repudiate the source of their own beliefs.

But, newly bereft of his sister, his main contact with Port-Royal, Blaise Pascal was in no mood to accept Jesuitical arguments from anyone, even his own collaborators. He and the other leaders of the Jansenist movement met together from time to time to discuss the situation, and their conversations grew ever more tense and bitter. After one of them, Gilberte reports, her brother was so shocked and filled with sorrow that he couldn't even speak. The sense of betrayal that he felt, coming especially as it was from Arnauld and Nicole, was strong enough that he felt overcome by sadness so great that he hadn't the strength to fight it.

Instead of remaining silent, however, Pascal sat down and wrote a short pamphlet entitled *Écrit sur la signature* (Commentary on the Signature). His main point was that Arnauld and company were wrong in assuming that the string of papal documents condemning the theology of Cornelis Jansen was merely a mistake, and that statements like "the doctrine of Jansenius with regard to the five propositions" were not a misunderstanding on the part of the papal theologians, and certainly not a misunderstanding on the part of the Jesuits. They were in fact a repudiation of the doctrine of St. Augustine on efficacious grace, for there was no difference between condemning Jansen and condemning St. Augustine and St. Paul himself.

Arnauld and Nicole responded that in fact Pascal was wrong about what was in the formulary, that it did not involve repudiation of St. Augustine or of efficacious grace, but only the repudiation of a straw man—that is, that it was a misunderstanding about what was actually in Jansen, Augustine's great interpreter. They told Pascal that he had unfairly condemned the pope and the theologians of the Sorbonne, for they were not evil men; they were not Jesuits or Molinists, but men of faith rowing their way through a complicated theological swamp. Why not sign the formulary, bow to the demands of the papacy, and save Port-Royal from the king? But Blaise was not interested in halfway measures, for, like Jacqueline, he was ready to start a revolution.

Actually, Blaise, who was not a theologian, and who could therefore see things a little more clearly, was more accurate in his assessment of the situation than Arnauld or Nicole. Of course, his neck was not on the line, whereas theirs were. No one required that he sign the formulary. He was quite correct in assuming that if the church rejected Jansen, who had accurately interpreted St. Augustine, it was essentially rejecting Augustine himself. "I conclude that those who sign the formulary, pledging only their faith while not formally excluding the teaching of Jansenius, have chosen the middle way which is abominable before God, contemptible before men, and entirely useless to those who are thus personally led astray."

This was the last battle of Blaise Pascal. In the coming months, his health declined precipitously. Gilberte and Florin Perier had moved to Paris just before Jacqueline's death—to care for their daughters, who had just been released from the school, and to see to the health of Blaise and Jacqueline. Sadly, they arrived in town on the day Jacqueline died. All they could do at that point was to sit by Blaise's bed and watch him slowly die. His fight with the leaders of the Jansenist movement had taken the last of his energy.

Meanwhile, the nuns of Port-Royal signed the second formulary on November 28 and 29, 1661, while appending a short profession of faith that Antoine Arnauld had written for them. That didn't suit the king or the Parisian vicars general, so they sent a representative to the convent and announced to the sisters that they would have to sign one more document, acknowledging the heretical nature of the five propositions found in the works of Jansen. But even this wasn't enough. On June 30, 1662, the King's Council ordered that all those required to sign the formulary agree to it "simply, and without restriction or addition." The Jansenists were finally caught. But not destroyed.

May God Never Abandon Me

We are usually convinced more easily by reasons we have found ourselves than by those which have occurred to others.

—BLAISE PASCAL, *Pensées* (1670)

It is the heart which perceives God and not the reason.

—BLAISE PASCAL, *Pensées*

Man is equally incapable of seeing the nothingness from which he emerges and the infinity in which he is engulfed.

—BLAISE PASCAL, *Pensées*

In the late summer of 1658, Blaise Pascal had a toothache. For most people, a toothache signifies little except the need to go to a dentist, but for Pascal, a toothache was like a dark storm cloud forming over a distant horizon. His old illnesses were returning. The pain was severe enough that it kept him awake at night, so instead of sitting and doing nothing in the late-night hours, Pascal decided to make use of his time by studying a complicated mathematical puzzle called the cycloid, or, in French, *la roulette*. Galileo had studied it, as had other great mathematicians, because its properties were peculiar and thus led to a set of difficult but interesting problems.

Essentially, if you take a point on a circle and draw a radius between that point and the center of the circle, and then roll the circle along a plane, as one would roll a wheel across the ground, the curve traced out by that point as it rolls looks like a series of arches and is called a cycloid. It was first studied by Nicholas of Cusa, who was eventually burned at the stake for heresy—but not for his work on the cycloid. Later, Galileo studied it and then gave the curve its name in 1599. When he tried to find the area underneath the curve, he failed to do so by using pure mathematics, so he eyeballed it by cutting pieces of metal to fit under the curve and then weighing the pieces of metal.

Père Mersenne took up the work in 1628 and tried to find the area under the curve by using integration but failed in his attempt, and passed the problem on to Roberval, who succeeded, though Descartes belittled his solution and then challenged him to construct a tangent to the cycloid. (Descartes had already succeeded at this.) Roberval failed, but Fermat soon succeeded. Descartes was both happy and sad. This is where Pascal entered the picture some years later, wide-awake with toothache. He developed a method for calculating the area of any segment under the cycloid, and then found how to calculate the center of gravity of any segment. After this, he showed how to discover the volume and surface area of the solid of rotation that is formed by rotating the cycloid around the x-axis.

Flushed from his success, he wanted to publish his findings but then had a sudden prick of conscience about his inherent worldliness and cancelled the entire affair. But then, as Gilberte pointed out, the duc de Roannez met with him and told him that he should publish his work because it would give extra credibility to his refutations of the unbelievers if he could also demonstrate the power of his logic by publicly solving some important problems in geometry, and thus give glory to God. Who could argue with that? Pascal took the duke's advice and published his work in October and December 1658. Afterward, also on the duke's advice, he announced a challenge to all mathematicians to submit their own solutions to the problems. Here, however, his scruples got the better of him and he issued the challenge under an assumed name, one Monsieur

Dettonville. According to the announcement, there would be monetary prizes for successful solutions. Sadly, the contest flopped, because only two people entered it, John Wallis and Antoine Lalouvère, and neither of them succeeded. A number of other mathematicians, including Pierre Fermat, Christopher Huygens, and Christopher Wren, the architect of St. Paul's Cathedral in London, passed their own solutions along without entering the contest; they probably didn't need the money. The whole thing was rather dodgy, however, because Pascal had already solved some of these issues himself, and he was also the one who had set up the board to judge the entries. Of all the backdoor solutions, Pascal most enjoyed Christopher Wren's method for calculating arc lengths, and he published it along with his own work. The only drawback to this new round of mathematical accomplishments, however, was that the purists at Port-Royal began to eye him with suspicion once again, and to chide him about being a man of the world.

All that ended as his illness deepened. In truth, he had been preparing for death ever since the night of his conversion. After the night of fire, friends and acquaintances of his often came around seeking his advice on religious matters, and they rarely went away unsatisfied. By 1658 and the return of the toothache, he had taken to wearing a belt with iron prickles under his clothes, next to his skin, in order to supercharge his penance. Any time he had a prideful thought, or felt pulled toward some diversion, he pushed on the girdle with his elbow, driving the points into his flesh, sharply reminding himself what his life was about. He wore that girdle until the day he died.

Sloth. The idle mind is the devil's workshop—that's what concerned him in his last years. He renounced all pleasure and all superfluity. Vain thoughts and desires were not for him, so he used his iron belt to keep himself on track. One by one, he removed the hangings in his room, turning his quarters into a bare monastic cell. He ate less than ever, and what he did eat was spare and simple. Only when under a doctor's orders did he force down anything heartier than broth. What's more, he recommended this same practice to others, not that many of them took him up on it. He often told people to abandon their desire for the best of everything, for things that were made by the best artisans out of the best material. Such things were useless, for Christ had called everyone to a holy poverty.

As he had in the past after an intense period of intellectual activity, Pascal took to his bed, for the old illnesses were bubbling back, but this time worse than ever. By the beginning of 1659, his digestive troubles had flared up, as had the headaches, along with the nausea and blurred vision. The symptoms of these headaches seem to indicate that they were migraines of a particularly nasty type, the kind of migraine that makes you sensitive to light, along with causing dizziness and vomitous nausea. Once again, the doctors circled around him giving him medicines that made him almost as sick as the diseases had, and with little salubrious effect. By March 1659, word got around among his circle of friends about the collapse of his health. His exhaustion was so great that he could barely rise from his bed, and it was a rare day when he could sit at his desk and organize a thought. The doctors trooped in and out, prescribing soup and, of all things, donkey's milk, which he forced down as a penance. Pain wracked him every day, and all the simple joys of his life flew from him.

Nevertheless, he had achieved a true detachment from the physical pleasures of the world, for, of the little money he still possessed after giving Jacqueline her dowry, he passed a great percentage on to the poor, either through charitable ministries in Clermont or by direct contact with the poor of Paris. Even in his reduced physical state, he often climbed out of bed and, following the example of St. Vincent de Paul, walked among the poorest of the poor in Paris and gave them what money he had.

Gilberte tells the following story: One day when he was coming home from Mass at the parish of Saint-Sulpice, he came upon a pretty little girl of fifteen who was passing among the people walking on the street, begging for money. He asked her how she had come to be begging, and she told him that she was from the country but that after her family had moved to the city, her father died. Her mother had been taken to the Hôtel-Dieu that very day. She was not expected to live. Blaise led her back to the rectory and provided the priest there with enough money to care for the girl and begged him to help her find a secure place in life off the streets of Paris. The priest agreed, and Pascal, before leaving, said that he would send someone to check on the girl. Then he disappeared into the evening, a stranger who never gave his name. Several days later, he sent a woman servant to the parish with more money. The priest asked her to tell him, for the sake of the girl, the name of her benefactor, but the woman responded that she had promised to keep his name a secret, so the girl never did find out who had been so kind to her that day.

Moreover, in the winter of 1662, still ferociously ill and close to death, he met a homeless family in the streets and invited them back to his house. He then assigned rooms to them to be their own and ordered them fed. And so they quickly joined the cobbled-together family that he had assembled for himself on the rue des Francs-Bourgeois.

Every day he drew a little closer to death, but somehow, like a meteor, he burned all the brighter. He had a brainstorm late one night that if he could only get a few investors to join him, he could set up a series of coaches that would take people from one part of the city to another for only five copper coins. This would allow the poor, who did not have their own coaches, to be able to move about more freely. Of course, the duc de Roannez was one of his investors, as were other members of Pascal's circle. In March 1662, the first coaches began to move between Porte Saint-Antoine and the Luxembourg Palace. Thus, Pascal's *carrosses à cinq sols* became the first public transportation system in the world. Of course, Blaise intended that any profits he made from this venture be given to the poor.

For some years, he had become increasingly isolated, and his illness only made his isolation worse. With Jacqueline in the convent, he had no one to look after his care as she had done, and although his servants were loyal, it wasn't the same. The loneliness, which had flattened him after Jacqueline entered Port-Royal, was still with him, but by now it had become a spirituality. "No one deserves to be loved," he told the gentlemen of the barns in one of his conferences. In his own mind, his isolation was no longer an empty, meaningless hunger for love; rather, it had been transmogrified into a purification of his soul, so that he could focus strictly on God. With the Pascal family scattered, and Blaise having no wife or children of his own, his servants had become his only family, and he was retreating even from them into the silence of his room, for, in bearing his pain in silence, he believed, he was preparing himself to see God. In these long months, sometime between 1659 and 1661, he wrote a heartfelt prayer, "A Prayer to Ask God to Make Good Use of Illness." One can almost hear his voice in it, crying from the depths:

> *Lord, your will is utterly good and sweet in everything, and you are filled*
> *with mercy. Not only the blessings but also the misfortunes that come upon*
> *your elect are the fruit of your mercy. Grant me the grace not to question*
> *you as a heathen would, even in the state of being to which your justice*
> *has reduced me.*[79]

Sadly, the dark Jansenist spirituality that so informed his life would not allow him to see the natural suffering and decline of the body as anything but a punishment for sin. Sadly, too, while Blaise's health gradually faded, the peace that Port-Royal enjoyed after the Miracle of the Thorn was shattered.

Meanwhile, Blaise lay dying, surrounded by all of his papers, his unfinished projects, especially his sheaves of notes on his apology for the Christian faith. Close to the end of 1662, after his sister and brother-in-law had transferred their family to Paris, Blaise surrendered his independence and moved in with them. His illnesses had become nearly intolerable, his pain a constant companion. On July 4, he asked his sister to send for the priest.

The Perier house lay in the parish of Saint-Étienne-du-Mont, near the monastery of St. Geneviève. The pastor there was a man named Paul Beurrier, who was known to be a good priest without ideological bias. Gilberte sent a note asking him to come by to hear her brother's confession, and one day in July he stopped by and administered the sacrament. Over the next six weeks, Père Beurrier visited on several occasions and held long conversations with the ailing Pascal, hearing his confession more than a few times. Pascal greatly appreciated these conversations, for they gave him comfort; Père Beurrier's simple spirituality and deep piety were a balm for his soul. For his own part, the priest remarked about how patient Pascal was in dealing with his infirmities. "He is so humble, he is submissive as an infant," he told Gilberte.[79]

By this time, news was getting around that Pascal had confessed his sins, and his friends began to appear at the Periers' doorstep to mournfully say good-bye. Soon, Pascal asked Père Beurrier to bring him the Eucharist, but his doctors protested, possibly out of the old superstition that once a sick person had received the last anointing, they were doomed to die. The doctors told Gilberte that the excitement of receiving the Eucharist might be too much for Blaise, and this irritated him, but he went along with them anyway, seeing how much of a disturbance his request was causing. After all, the followers of the abbé de Saint-Cyran took Communion only rarely, fearing that their sinfulness might pollute the sacrament, and so many of them held off receiving the Eucharist until just before they died.

In spite of everything his doctors tried, Pascal's condition quickly deteriorated. His life was burning away, and although he had his good days when he could visit with his friends, those days had become increasingly rare. His most poignant visit was with Antoine Arnauld, who came dressed in disguise for fear of the king's secret police. The two men talked over their differences and came to a final reconciliation. By August 3, Blaise had settled all of his affairs and signed his will. Eleven days later, he had a sudden attack of dizziness and a massive headache. After three days, he fell into convulsions. On the night of August 17, the terrified Gilberte sent for Père Beurrier, who rushed to Pascal's bedside and administered viaticum, the Eucharist of the journey, given to those who are about to die. Afterward he gave him extreme unction, which is now called the anointing of the sick. Pascal received the sacrament with tears in his eyes, and when the priest took the ciborium and traced the sign of the cross in the air with it as a final blessing, Pascal cried out, "May God never abandon me!" Soon after, he fell into a coma, and at one o'clock in the morning on Saturday, August 19, 1662, he died in his bed. He was thirty-nine years old.

But this wasn't the end of the story. On Monday, three days later, Père Beurrier offered Pascal's funeral Mass at the parish church of Saint-Étienne-du-Mont, and afterward they laid his body in a tomb inside the church, behind the high altar of the Lady Chapel. Eighteen months later, the Perier family erected a small plaque on the wall near his tomb. The plaque merely said that he had spent his last days "meditating on the law of God." The war against Jansenism was in a fever pitch, however, and someone quickly reported to the new archbishop of Paris, Hardouin de Beaumont de Péréfixe, that the plaque was an obscenity, since Pascal had died a heretic who had refused the sacraments. They demanded that his body be removed from the church.

The archbishop, a consummate politician, realized that he had an opportunity now to gain a few points against the Jansenists, so he called in Père Beurrier and asked him if indeed Pascal had died the way his enemies had claimed. By this time, Pascal had been discovered as the author of the *Provincial Letters,* and so there were those who were looking for any way to defame his memory. Père Beurrier reported that Pascal's enemies were in error, for he himself could attest to the fact that Pascal had received the sacraments before he died and had died in the good graces of the church.

The archbishop asked the priest to write down everything that he could remember about Pascal's death, and asked him to give special attention to the fact that Pascal had received the sacraments and the fact that he had denied Jansenism before he died. After some arm-twisting, Père Beurrier did so, writing that "he had formerly belonged to the party of the gentlemen of Port-Royal, but that for two years he had been estranged from them, because he discerned that they went too far in matters related to the doctrine of grace."[80] The archbishop made sure that this account became public.

Pierre Nicole, however, responded to the priest's account by saying that, far from estranging himself from the Jansenist movement, Pascal had chided the other leaders of the movement for their willingness to sign the formulary, and had encouraged them to resist what he considered an unfair and unjust demand. Both Gilberte and Florin Perier agreed with Nicole's account. After this, the controversy simply blew away with the shifting winds. Pope Alexander VII died, and his successor, Clement IX, negotiated a peace between the Jansenist movement and the rest of the Catholic Church. This became known as the Clementine Peace, but it didn't last more than ten years. After that, the struggle started up once again and continued until the French Revolution erupted and set fire to the world.

Oracles, Dicing, and Schrödinger's Cat

The free man owns himself. He can damage himself with either eating or drinking; he can ruin himself with gambling. If he does he is certainly a damn fool, and he might possibly be a damned soul; but if he may not, he is not a free man any more than a dog.

—G. K. Chesterton

If scientific reasoning were limited to the logical processes of arithmetic, we should not get very far in our understanding of the physical world. One might as well attempt to grasp the game of poker entirely by the use of the mathematics of probability.

—Vannevar Bush (1890–1974)

I am the abbot of Cockaine
and my congregation are all drunks,
and I wish to be in the order of Decius,
and whoever looks for me at tavern in the morning,
after Vespers he will leave naked,
and thus stripped of his clothes
he will call out:
Woe! Woe!
what have you done, vilest Fate?
the joys of my life
you have taken all away!

—Carmina Burana

Sometimes, the primeval Greeks discovered a new valley, a new stand of forest that would speak to them. Often, there was a cave, or a spring pooling, with a stream bleeding off the overflow; often, it was on a high place, close to the gods, or by an oak tree struck by lightning, like scribbles on the earth saying "Zeus was here." Something about the light, about the shape of the hills, about the way a stream bubbled through the trees made them feel the presence. They could not name this feeling, but there was always something uncanny about it, something numinous, and it frightened them and drew them in at the same time. "Surely a god must live in this place," they said, and ordered games to tease out the god's name so that they could build a temple there. There were racing games, and wrestling games, and archery contests, and discus throws. There were sacrifices of bulls and goats to the unnamed god. At the height of the festival, the priests, after purifying themselves, took sticks with letters or numbers carved into them, or took astragali, the knucklebones of sheep with numbers carved on the uneven faces, and would ask the god questions:

"What is your name?"

"What do you command?"

"How can we worship you?"

There was a throw, a reading, an interpretation, another throw, and by such fits and starts the gods revealed themselves. "It is Athena who lives here," the priests said after multiple attempts in one such place, and the people built a temple to her, and initiated a cult to her name. Around that cult grew a city that survived through the ages.

Human life has the smell of oddness about it. We neither exist nor not exist; we stand somewhere in between. If we existed, and that alone, we would never die, and if we did not exist, and that alone, we would never be. But neither of these are our lot; we are suspended between such purities, and so we live out our lives like ghosts, caught between the worlds, trying hard to forget what we know by smell, what we know by the play of light on the eye—that life is evanescent, a dream, a flash of sudden glory. Beyond that, no one knows, for no one ever returns. We can either ignore this truth, pretend it isn't so and go on diverting ourselves, or we can face it, hard cold fact that it is. Some people who do face it brave death and jump out of airplanes; others go to war; others enter the monastery.

And some people gamble. Like the Greeks, who were insatiable gamblers, they risk great things on the throw of the dice. They ask questions of the shadows: Will I be rich? Will I find happiness? For the Greeks, the goddess who answered was called Tyche, the goddess of fortune, the goddess of luck, success, and chance. Like Demeter, who granted bountiful harvests, Tyche granted sudden fortunes, but then her partner, her other face, Nemesis, took them away. And then the people complained. "The lowest possible specimen of humanity, one who as the victim of Fortuna (Fortune) [Tyche] has lost status, inheritance and security, is a man so disreputable that nowhere in the world can he find an equal in wretchedness."[81]

Somehow, however, there is always a god in the world, hidden just out of sight, in the shadows, in that same play of light, in the feeling that something right nearby is sacred. Anyone who attends to the world will sense this, sooner or later. Even if that god has no name, the feeling is there, and is not limited to believers. Even those who do not believe in the god feel it, for belief has nothing to do with it; the feeling comes before the believing. It is primal, emerging from the ancient Olduvai strata of the mind, and it can be ignored or explained away, but it cannot be avoided. The only way to never feel it is to deaden your senses and sleep, and never really live.

But like the Greeks, we have games. We gamble because it is more than a game. It is a practice run on life, a ritual reenactment of the truth of human existence—that we are all gamblers, and that we face the mystery, that we throw the dice and ask the questions. You interrogate a gambler about his behavior, and he will inevitably say that it is fun. It is a sport, and cannot be analyzed beyond that. But that is not what really happens. In the act of gambling, the gambler invariably creates a narrative that has little to do with reality. Before each throw, gamblers, like the saints, are winners, and in the winning they are translated to some new emotional place.[82] It doesn't matter if you point out that the house odds are against them, and that the story they tell themselves down deep, where no one can see it, is an opium dream, and that they run toward potential ruin like berserkers. Gambling, therefore, is a brave act, but brave in the way that a masochist is brave. It is also an act of stupendous hope. As in sex, a gambler's pleasure is brief, lasting between the time the dice are thrown and the time the numbers are up. All the rest is devastation or afterglow. But we don't live for the afterglow; we live for that moment, right then as the dice are tumbling and the world is at hazard. That instant, artificial as it is, is like a short burst of mysticism; the gambler stands fixed, not breathing, waiting on God for a word. Like sex, it is a single instant of eternity.

And that is why so many people get hooked.

Blaise Pascal ruined all of this by creating a new myth, the myth that one could have some control over the future through calculation. Cast adrift in the world, he was suddenly his own man, and with his friends, he attended salons and happy parties, standing on the sideline watching the gamblers try to keep their cool. It was a gentlemen's game, a gentlemen's version of combat, an act of honor, wrestling with Fortune and facing the wall of shadows. Did he understand then what was at stake? Possibly. Certainly, he was guilt-ridden about his life, for he had said so to Jacqueline on many long visits. He was lonely, and diversions helped him stem the long, sad hours.

He filled his days with conversation, with his researches into mathematics and physics, and eventually made new friends. Blaise and the pious and intractable duc de Roannez became close, and Blaise even sort of fell in love with the duke's sister, though he might not have called it that, and though he treated her as if he were her father confessor. He was certainly attached to her, and wrote pious letters to her about salvation, which always turns a girl's head. Through those friends he made associations with other, less savory friends, like the chevalier de Méré, the inveterate gambler and onetime saloon keeper, and others. The duke and his sister shared Pascal's spirituality; the chevalier de Méré did not. However, all of them seemed to find the young man, sometimes arrogant, sometimes quiet, sometimes brooding, an amiable companion, a true *honnête homme,* a Christian gentleman of integrity. But this was not the real Pascal, who found himself gradually sinking from one depression into another.

This was a deeply creative time for Pascal, however, as the ideas that later became the *Pensées,* especially the wager, precipitated out of the air of the salons. Pascal observed diversion in its elemental form, where knights hacked at each other with dice and betting chits and debts of honor. Supposedly, everyone expected the aristocrats to hide their devastation, for letting it out would have been the gambler's equivalent of running before the enemy. But it happened, more often than not. Louis XIV's mistress Athenaïs was notorious for throwing epic fits, ecstatic when she won and theatrical when she did not.

For Pascal the Jansenist, all of this was too much a part of the world, where the damned tarried for a short while before falling headlong into the flames. Still, his experiences as a kibitzer, and perhaps even as a gambler, were not devoid of religion, because Pascal perceived the action of God even in the throwing of dice. How else could he have come up with a gambler's argument for faith, his famous wager? There were two vectors of experience in Pascal's world, vectors that for inattentive people would have remained unconnected but that for a creative mind might well come together in new and interesting ways. First of all, in this time of his life, Pascal was a man of the world, living up to the expectations of his class and living the life of an intelligent gentleman. This meant parties; this meant diversion; this meant gambling. It is unlikely, given his health, that he engaged in any dangerous liaisons, nor would his conscience have allowed it. So instead, he invented the new mathematics of probability, the calculation of expectations, a mathematics for probing the mysterious heart of Fortune.

Theologians have long discussed a category of thought called Mystery— that is, the ultimate truth of the universe, which is beyond our rational powers. Pascal accepted this idea as a given. Modern physics has rediscovered that category, obliquely at least, by noting that there are dimensions of the physical universe that we cannot measure, and by noting that in that limitation, we cannot know but only suppose. This is a limitation on reason, *ratio* in Latin; for reason and measurement are siblings, if not nearly the same thing. But oddly enough, while we talk a great deal about knowledge and reason, it is Mystery that moves us. Mystery seduces us. It is the essential oil of our best joys, from eroticism to mysticism, and of our worst terrors, from atomic war to hell. What you do not know, or cannot know, drives the heart as well as the mind, insinuating its odor into your entire self as you attempt to penetrate what by definition cannot be penetrated. Mystery is the heart of ancient religion—the feeling of dread in the presence of a dead body, the feeling of awe before the thunderstorm, the feeling of immensity under the night sky. It is the oceanic, the sublime, the greatest of fears and the greatest of wonders, and we are desperate in front of it. It is like walking into a pitch-black room and fumbling on the wall for the light, while from just over your right shoulder you hear the sound of heavy breathing.

Classical science has worked hard to dispel this category, waving it off as superstition, and assuring the public that with time science will explain it all. Of course, if scientists actually managed to do that, people would rise up and kill them. This is one reason why there has been such an emotional hubbub between science and religion. For, if religious people were to allow the scientists to explain it all in the squiggles of mathematics, it would take the joy as well as the terror from the world, and we so love both. If scientists, on the other hand, were to open their science to any concept that had the incense smell of religion about it, then, they fear, and perhaps rightly so, that they would be letting Mystery in by the back door, and the quest to explain it all would fail.

On the other hand, the explanations that scientists have handed down to the citizens have been chronically disappointing, especially after the nineteenth century, when scientific naturalism spilled over the causeway of method into metaphysics. As a species, we went out looking for the mind of God, and instead we discovered dust and gas. So once again, we human beings are more moved by what we do not know than by what we do know, which may be endemic to the species and a fairer assessment of the human psyche than any other. Why else would we have left Olduvai Gorge?

Every day is a gamble, and just by climbing out of your bed and leaving your front door, you become a gamester. You calculate the odds, and you take the risks. This is what gives life its spice, for there is always a risk. Each morning over coffee, after the newspaper, you throw the dice. You start your day in uncertainty, because no one really knows what will happen from moment to moment. The future doesn't really exist, except as a potential, as a daydream, as an intimation conjured from the present. The past is a memory that over time becomes a dream, and the present bubbles along, ephemeral as a thought, an everlasting mutation, making us feel as if the future were here, now, close enough to touch, but still elusive as smoke.

So we stand before the Mystery of this moment, terrified and awestruck as this moment dies and is reborn as the next. The mysteries of time and death are at the center of the human condition, and it has been so since we first achieved consciousness. This is the beginning place, Uncle Wiggly's first hop, the place out of which all science, all religion, all literature, all philosophy, all art and music are born. Fortune and misfortune pour toward us from the middle of the thunderstorm, and we interpret their coming in the best way we can. For some people, the source of that coming is a divine figure, God, who lives in the heart of the Mystery and is Mystery itself. Any attempts to plumb the personality of that divinity by reason are called theology. For others, these things come of their own, pure chance, and cannot be tracked any further. And so we have grace and luck, two concepts that were at the heart of Pascal's thought.

In a sense, grace and luck are not all that different. Both fall out of the surrounding mystery. The fact that God gives grace to some and not to others, as Augustine would hold, is almost a matter of luck. Why God chooses to love one and not another is weird beyond weird.[83] The fact that the dice would come up sevens this one time when you need it so badly is a mystery of chance, and what is chance but a word that says, "I don't know." Who is Fortune that she gives her gifts to some and not to others?

But this strange idea—chance, luck—has become the foundation for an entirely new physics.

From the point of view of science, Pascal's probability theory has set the stage for a new metaphysics, embodied in quantum mechanics, which places the most fundamental level of matter, the subatomic particles, into a shadow land between existence and nonexistence. They exist as probability clouds rather than as things, and they can never be tracked, never precisely measured. The world gets stranger by the day.

Quantum physics is the physics of the unbelievably small. Newton's physics is good for describing the everyday, from baseballs to rockets to the moon. It can describe the orbits of planets and of spaceships, the bouncing of billiard balls and the arcs of cannon shells. But something unusual happens when things get small. The rules seem to mutate a bit even at the level of grains of sand. But down past the molecule, as electrons swirl like bees around the nuclei of atoms, the normal laws of everyday physics seem to break down and become something else. The old division between being and nonbeing breaks down with them, as subatomic particles pop into being and then pop out again. Electrons exist not in this place or that, but in those probability clouds that surround the nucleus and spin so fast you can almost hear them hum. The world at this level, then, is not an existing one but a probable one.

At the bottom of quantum physics is the discovery that light has the qualities both of a particle and of a wave, a wave/particle duality. In the blackbody experiments of Max Planck, in the photoelectric-effect studies that got Einstein his Nobel Prize, and in the scattering of photons that Compton noticed, light seemed to act like a particle. But then, anyone who has had high school physics knows that light is subject to refraction and will produce interference patterns like waves on a lake, and in that it acts like a wave. Therefore, we cannot explain light without both pictures. Once we got rid of Christian Huygens's "lumeniferous aether," the undetectable medium that light is supposed to travel through as it moves through a vacuum, the vacuum championed by Pascal, the nature of light took on two faces.

This is where Werner Heisenberg comes into the picture. During one of the conferences at the Copenhagen Institute, he asked what would happen if we wanted to measure both the position and the momentum of a specific particle. In order for us to do this, we must be able to "see" the particle, so that we can measure it. So, to see it, we have to shine a light on it. The light has a wavelength λ. But in order to actually see clearly, the smaller the object to be observed, the shorter the wavelength λ will have to be, until at the level of electrons, we will have to use such powerful gamma rays to see it that we will change it by our seeing. In this way, Heisenberg realized (actually following the philosophy of Immanuel Kant), we will never see the electron as it truly is, but only as it is once we've changed it by our seeing. This gives an *uncertainty* to the particle's position ($\Delta x \approx \lambda$).

This means that the change in x caused by our seeing is approximately, but uncertainly, equal to the wavelength of the light that we are using to see it with. Now, this applies if we see light as a wave. If we see it as a particle, then we can say that the light we use to see it with gives up some of its momentum to the particles when it illuminates it. How much it gives up is unknown, and unknowable, because it is immeasurable. So, in this case, the change in momentum (h) is described by the following formula:

$$\Delta x \, \Delta p \approx h$$

We must note here that this uncertainty is not something that comes from the inadequacy of our technology or of our methods, but is so in principle. You cannot precisely measure both the position and the momentum of a subatomic particle. If you try to get a more accurate fix on one, the other will go askew, and contrariwise.

Now things really get strange. The Schrödinger equation, which spins off of Heisenberg, demonstrates how the basic nature of things is probabilistic and not classically real. The thought experiment that he came up with took Heisenberg's uncertainty principle and generalized it by describing certain wave functions about whose outcomes, given a large number of outcomes, the Schrödinger equation will be able to make predictions. Most people have heard about his famous thought experiment about the cat placed in a box with a vial of cyanide gas that would break or not break depending upon whether a specific sensor is hit by a subatomic particle. The chances of its being hit are fifty-fifty. For Schrödinger, as long as the box remains closed and no one looks inside, the wave function of the particle and the state of the sensor remain undecided, and the cat is both alive and dead. It is only when someone opens the box that one wave function collapses and the cat turns up one way or the other. We all cheer for the cat to make it.

Note the term "large number of outcomes" in the last paragraph. This is the part that has a direct bearing on the letters of Pascal and Fermat. In his letter on the problem of expectations, Pascal was aware that as the number of throws of the dice increased, an effect was produced on the expectations that the players could legitimately hold in relation to the game. Later practitioners of the arts of probability have studied this effect and have used it to send boats deep into the continent of this new mathematical world, into the land of big numbers.

One of the things they discovered is the so-called law of averages, or Bernoulli's law. Many gamblers mistakenly think that this law means that everything will average out, that if over ten tosses of a coin you have six heads and four tails, eventually—over, say, a hundred or a thousand throws—there will be a corrective and the player will begin to get more tails to make up the difference, so that eventually the number of heads and tails will be the same. In other words, in the long run, the chances will even out. But this isn't how it really works, though it's a nice enough idea to have its own name, the "gambler's fallacy," or the "maturity of chances." It has lost a lot of people a lot of money.

To understand the way the "law of big numbers" works, you have to understand two things: First, every time you toss a coin or throw a die, the probability of a single outcome is always the same. The probability of getting a heads for each toss is fifty-fifty, while the probability of getting a six on the toss of a single die is one in six, while the probability of getting double sixes with two dice is one in thirty-six. This means that every toss is a fresh start, and the uncertainty of the outcome is just as great as in the first toss. This takes care of the gambler's fallacy, because there is no guiding hand making sure that the tosses even out, and at every moment while the coin is spinning, even the powerful mathematics of probability can't make a prediction of what will happen next.

Let us take the example of a coin toss. The "law of averages," or the "law of big numbers," says that as the number of times we toss the coin grows, the number of heads we throw will grow closer, in proportion, to half the number of total throws. Therefore, as the number of tosses grows, the probability grows that the percentage of heads (or tails) will get closer to fifty. But the actual number of heads thrown will get larger, too, just as will the percentage. So, we flip a coin a hundred times and get sixty heads and forty tails, which is 60 percent to 40 percent. The gambler's fallacy would argue that somewhere in there, ten more tails will show up to make up the difference. But you keep flipping and writing down the result. You get to a thousand flips, and the difference is now 55 percent to 45 percent, but the difference in the number of throws is no longer a mere twenty throws; it is now a hundred throws. This is the Law of Big Numbers, and odd things not only happen but become commonplace as the numbers get larger.

Two people meet each other on the streets of New York just after 9/11, and after further discussion find that their grandparents once had a similar encounter on the streets of London after the Blitz. Then they find out that their grandparents' grandparents had a similar encounter on the streets of Paris after the First World War. What are the possibilities of that? With a big enough human population, it would be a dawdle. And so, in a big enough universe, anything can happen. The only problem for really improbable things is whether the universe is big enough. And with a big enough universe, say some, you don't need anything else.

At this point in Western intellectual history, the old division between materialism and religion seems unbridgeable. Religious people believe ever more deeply in the mystery that the Greeks first experienced in the glades and forests, the running streams with intimations of divinity haunting the field. As such, they are more directly tied to the long drive upward of the human race, and find ever more creative ways to express the new cosmology in theological terms. The old ways are never forgotten; they just find new ways to get into the papers. The materialists, however, are still holding to the notion that we would be better off without all that supernatural folderol, and so they spin ever more delightful theories to make that happen.

The cultural landscape still seems to be divided into two strategic types—the climbers and the sprawlers. The climbers live vertically, and find ever more wondrous and exotic mysteries in the everyday world, mysteries that somehow intimate God, just as the forests and glades once did. The sprawlers live horizontally. They find satisfaction in the idea of many chances and of big numbers. Anything can happen in the universe if the numbers are big enough. The vastly improbable becomes probable; the uncanny becomes ordinary. Eventually, with big enough numbers, those ten thousand monkeys could type out not only Shakespeare, but Milton, too, and half the Bible. Big numbers explain the coming of life and the evolution of intelligence. Of course, the old rhetorical project lurks behind the big numbers. Why have a God when you can have googolplexes? With big enough numbers, anything can happen.

The ultimate sprawler's theory is the multiple universes of John Archibald Wheeler. We can avoid the whole question of Providence, even in light of the big bang, by inventing a zillion zillion universes, separate from ours, where every conceivable quantum state can have its day. In this way, my dog who was hit by the car is alive in some other universe. The fact that such universes have no more empirical evidence for their existence than does God, and are no more falsifiable than is God's existence, doesn't seem to matter. In fact, the whole thing might balance on the edge of Occam's razor. But which is the simpler explanation? If you have big enough numbers, you don't need God, and that is the heart of it. But are multiple universes any simpler an explanation than God? It seems finally to come down to choice, perhaps even to the Two Standards: people who believe in God do so because they want to; people who don't believe don't because they want to. Almost makes one think of efficacious grace.

NOTES

1. John R. Cole, *Pascal: The Man and His Two Loves* (New York: New York Univ. Press, 1995), 24.
2. Cole, *Pascal*, 19.
3. Blaise Pascal, *Pensées and Other Writings*, trans. Honor Levi, ed. Anthony Levi (Oxford: Oxford Univ. Press, 1995), 8.
4. Let us be clear here. It is my contention, with Shakespeare, that in the Reformation/Counter-Reformation period, "all are punished." Albrecht Wallenstein may have been ruthless, but he was no more so than Protestant generals of the same period. Too many myths about good guys and bad guys still make the circuit. For every massacre on St. Bartholomew's Day committed by the Catholics, there was an equally horrible one committed by Oliver Cromwell and others like him in Ireland, or America, or Germany, or Bohemia. But not one of these men committed crimes to equal what was done by post-Christian butchers like Hitler, Stalin, and Mao.
5. Pascal, *Pensées*, 66.
6. Some later scholars have argued that Blaise may have had a secret copy of Euclid hidden away, which is possible. But if he had, why invent his own definitions? Was Blaise hoping that his father would catch him, as eventually he did, or was he doing this all for his own amusement, as Gilberte indicated in her *Vie de M. Pascal*? Who can say?
7. Fermat's last theorem is that the equation $x^n + y^n = z^n$ has no nonzero integer solutions for x, y, and z when n is not equal to 2. In a marginal note, Fermat wrote: "I have discovered a truly remarkable proof, which this margin is too small to contain." The problem was that Fermat's own writing was sloppy and disjointed, so much so that he could never get it printed. Fermat never mentioned this proof again, and mathematicians have been trying to rediscover his remarkable proof ever since.
8. Nor could you have told it to the two queens, who remained unbowed. After all, Anne was a Hapsburg and deeply involved in the Spanish intrigues at court. The Hapsburgs were, after all, the *über*-Catholics, the ones holding the line against all those dirty Calvinist heretics, who, to the Spanish mind, were good, barring conversion, only for fuel. Marie de Médicis, on the other hand, was an Italian and wanted to convert the French court into something more Tuscan.

   In 1629, Richelieu had already conquered the Huguenots (the French Calvinists), and his power over the government seemed complete. At first, he had tried to conciliate between the Protestants and Catholics, and was prepared to be tolerant, which irritated the Spanish faction no end. Religion was the most divisive issue of the time, and extreme elements on both sides had wrecked his plans. The French Calvinists had built their own army and fortified castles. Then the Huguenots made the mistake of involving themselves in an attempt to pressure England to declare war on France. And so Richelieu declared war on the Huguenots, and led the army himself to besiege the castle of La Rochelle, where the Protestants were holed up. Meanwhile, Spain tried to solidify its power in Italy by attacking France's allies in the north. Richelieu's secret police caught wind of certain secret communications going on between Queen Anne and her brother, the king of Spain. Marie de Médicis, on the other hand, still stung by Richelieu's departure from her camp, and possibly suffering from unrequited love, allowed herself to be swayed by the radical Catholics to get rid of the more evenhanded cardinal, because they said he was too tolerant of the Calvinists.

So the two queens drafted their plans. Mortal enemies up until this time, they reconciled long enough to try to get rid of Richelieu, whom they both despised more than they did each other. Marie de Médicis, the Queen Mother, who could swear like a dockworker and was heartless to boot, had led a hard life. Her uncle, Ferdinando I de' Medici, bought the crown of France for her for six hundred thousand crowns and married her to Henri IV. But Henri already had a mistress in place, a vicious mistress, Henriette d'Entraigues, who claimed to be the queen herself because of a foolish document Henri had signed when he was particularly smitten, and maybe a little drunk, and who did not want to share him, least of all with this foreigner from Italy. Henriette instigated open war between the king and his queen, though the queen still managed to give Henri a brace of sons. There were shouting matches and haughty exchanges until the king, happily for Marie, was stabbed to death by an assassin on the rue de la Ferronnerie. By the end of the day, Marie was the regent of France, and Henriette d'Entraigues was in a lot of trouble. One evening after Richelieu's return from war, on what was later called the Day of Dupes, Marie met with Louis privately in her chambers. Fearing spies, she set her guard at the door, and when Richelieu got wind of the meeting, he tried to bully the guards into letting him pass but was rebuffed. So, fearing for his life, the all-powerful cardinal snuck into the Queen Mother's apartment through the side chapel and appeared before the king and his mother. Marie, raging, told Louis that he must choose between Richelieu and her—one of them would have to go. The king agreed to fire his first minister, and Richelieu was certain that his life was forfeit, but within a week, Marie de Medicis was running for her life toward the Dutch border.

9. Susan Griffin, *The Book of the Courtesans: A Catalogue of Their Virtues* (New York: Broadway, 2001).
10. Thomas M. Kavanagh, *Dice, Cards, Wheels: A Different History of French Culture* (Philadelphia: Univ. of Pennsylvania Press, 2005), 14.
11. Kavanagh, *Dice, Cards, Wheels,* 40.
12. Latin text from which this translation was made can be found at http://www.fh-augsburg.de/harsch/chronological/Lspost 13/CarminaBurana/bu-caro.1html.
13. John's Gospel is rife with this kind of thinking. Nicodemus cannot understand what Jesus is saying, whereas the woman at the well understands after only a short conversation. The blind man sees, but the priests of the temple are blind. Though such passages may be anti-Semitic at some levels, they are primarily about a fundamental social experience—that some people hear the truth of religion, while others are deaf to it.
14. *Quatrains du déiste,* stanzas 1–3. Cited in Antoine Adam, *Les libertins au XVIIe siècle,* 90. Quoted in David Wetsel, *Pascal and Disbelief: Catechesis and Conversion in the "Pensées"* (Washington, DC: Catholic Univ. of America Press, 1994), 97–98.
15. Wetsel, *Pascal and Disbelief,* 99.

16. Marvin O'Connell, *Blaise Pascal: Reasons of the Heart* (Grand Rapids, MI: Eerdmans, 1997), 23.

17. William R. Shea, *Designing Experiments and Games of Chance: The Unconventional Science of Blaise Pascal* (Canton, MA: Science History Publications, 2003), 12.

18. Blaise Pascal, "Letter to the Chancellor About the Adding Machine," in *Pascal: Selections,* ed. Richard H. Popkin (New York: Scribner, 1989), 21.

19. This was also true of Harrison's clock, which he invented to solve the longitude problem. Even after the clock was invented, many sea captains sailed by astronomical tables rather than the clock because the cost was prohibitive.

20. Shea, *Designing Experiments,* 12.

21. Pierre de Bérulle, "Discours de l'état et des grandeurs de Jésus," *Œuvres complètes* (Paris: Migne, 1856), 161. English translation in Anne M. Minton, "The Spirituality of Bérulle: A New Look," *Spirituality Today* 36, no. 3 (Fall 1984): 210–19.

22. In my own recounting of the history of the vacuum, I am deeply indebted to William Shea's detailed account of all the comings and goings of research on this subject. If you want to find out more about this subject, see Shea's *Designing Experiments,* 17–18.

23. Shea, *Designing Experiments,* 42.

24. Popkin, ed., *Pascal: Selections,* 35.

25. Emile Cailliet and John C. Blankenagel, trans., *Great Shorter Works of Pascal* (Westport, CT: Greenwood Press, 1948), 44.

26. Pascal's reply to Père Noël, in Cailliet and Blankenagel, *Great Shorter Works of Pascal,* 49.

27. Jacqueline Pascal, "A Memoir of Mère Marie Angélique by Soeur Jacqueline de Sainte Euphémie Pascal," in *A Rule for Children and Other Writings,* ed. and trans. John J. Conley, S.J. (Chicago: Univ. of Chicago Press, 2003), 132.

28. Jacqueline Pascal, "Memoir," 133.

29. Pascal to Perier, 15 November 1647, *Œuvres de Pascal* 2:680. Translated in Popkin, *Pascal: Selections,* 44.

30. *Œuvres completes de Pascal* 2:682–84.

31. Shea, *Designing Experiments,* 115.

32. O'Connell, *Blaise Pascal,* 55.

33. O'Connell, *Blaise Pascal,* 53.

34. Jacqueline Pascal, "Memoir," 125.

35. At least according to Cardinal de Retz, who was no great friend of Richelieu's. If Retz is correct, then Richelieu's service, which so dominated the court, must have been bitter to the king, who wanted to be his own man all his life. In this account, I am relying on the memoirs of Madame de Motteville and of Cardinal de Retz for many of the details about the Fronde. Motteville was a lady-in-waiting to Queen Anne, and praises her virtues while ignoring her faults. Retz was an ambitious man ill suited to the church by his own admission. We have to take what both of them say with more than a little skepticism, because both of them were partisans of one faction or other, and both were ferociously ambitious people who used their memoirs to make political points and to excuse their lives. See Jean François Paul de Gondi, *Memoirs of Jean François Paul de Gondi, Cardinal de Retz* (Boston: L.C. Page and Company, 1899); and Françoise de Motteville, *Chronique de la Fronde* (Paris: Mercure de France, 2003).

36. Gondi, *Memoirs,* 145.

37. The French word *fronde* is an odd word, meaning "sling." During the uprising, the people used slings to bombard the windows of Cardinal Mazarin's supporters with stones. This

uprising began as an attempt by the general populace to be heard, but it soon degenerated into factions and then exploded into an attempted coup d'état. At issue were the tax policies of the dead Cardinal Richelieu.

38. Cailliet and Blankenagel, *Great Shorter Works of Pascal*, 87.

39. This was a perverted version of the Two Great Commandments in the Gospels: that you should love the Lord your God with your whole heart, your whole soul, your whole mind, your whole strength; and that you should love your neighbor as yourself. It was perverted because Pascal, following the Augustinian tradition, believed that sin subverted the individual's love of God. Moreover, the love of self was not the measure by which we come to understand charity. For the Augustinian, one should love the neighbor and hate the self, thus denying the intimate connection between self-love and charity that the Gospels called for.

40. O'Connell, *Blaise Pascal*, 81.

41. O'Connell, *Blaise Pascal*, 81–82.

42. *Œuvres de Pascal* 2:976. Quoted at length in Cole, *Pascal*. Cole's rendition of the struggle in the Pascal family is one of the best I've ever seen.

43. *Œuvres de Pascal* 2:749. Quoted in Cole, *Pascal*, 83.

44. Jacqueline Pascal to Blaise, May 7–9, 1652, in Jacqueline Pascal, "Memoir," 140.

45. O'Connell, *Blaise Pascal*, 83.

46. O'Connell, *Blaise Pascal*, 82.

47. O'Connell, *Blaise Pascal*, 7.

48. Here is the math: $1-(\frac{5}{6})^4 \approx 0.5177$ is the probability of rolling one six in four rolls. The probability of getting two sixes in twenty-four rolls is $1-(\frac{25}{36})^{24} \approx 0.4914$, which is slightly lower than the probability of getting one six. The chevalier was correct in his observation, but imprudent in his betting.

49. O'Connell, *Blaise Pascal*, 89.

50. *Œuvres de Pascal* 2:1138. Translation taken from Shea, *Designing Experiments*, 264.

51. *Œuvres de Pascal* 2:1138. Translation taken from Shea, *Designing Experiments*, 264.

52. Shea, *Designing Experiments*, 255–56.

53. The mathematics are far too involved to get into here. Those who are interested in such things should consult William R. Shea's *Designing Experiments and Games of Chance*. Shea's account is lucid and entertaining, and better yet, readable.

54. Shea, *Designing Experiments*, 284.

55. *Œuvres de Pascal*, 2:1034–35. Translation quoted from Shea, *Designing Experiments*, 288.

56. Cailliet and Blankenagel, *Great Shorter Works of Pascal*, 118.

57. Cailliet and Blankenagel, *Great Shorter Works of Pascal*, 118.

58. Cailliet and Blankenagel, *Great Shorter Works of Pascal*, 118.

59. Cailliet and Blankenagel, *Great Shorter Works of Pascal*, 118.

60. The French version can be found on the Internet at http://www.users.csbsju.edu/~eknuth/pascal.html. An English translation is also provided there, but I present my own here.

61. Conversation with M. de Saci (1655), cited in Popkin, *Pascal: Selections*, 79. This was not verbatim of the conversation, but a report written later by Père Singlin's secretary.

62. O'Connell, *Blaise Pascal*, 105.

63. Cailliet and Blankenagel, *Great Shorter Works of Pascal*, 122.

64. Popkin, *Pascal: Selections*, 82.

65. I love these old church documents. Who can come up with words like *contumelious*?

66. O'Connell, *Blaise Pascal*, 128.

67. Blaise Pascal, *Pensées et provinciales choisies,* trans. Stanley Applebaum (Mineola, NY: Dover, 2004).

68. Pascal, *Pensées et provinciales choisies,* 8.

69. Blaise Pascal, *Provincial Letters,* trans. Thomas M'Crie (Eugene, OR: Wipf & Stock, 1997), 16.

70. Blaise Pascal, *Provincial Letters,* 150.

71. Quoted in O'Connell, *Blaise Pascal*, 136.

72. O'Connell, *Blaise Pascal*, 166.

73. Pascal, *Pensées and Other Writings,* 67.

74. I have translated the argument in full:

Now let's speak in accordance with our natural understanding.

If there is a God, he would be unfathomable, and infinitely so, since, having neither parts nor boundaries, he would exist in a way that we could not even imagine. Thus, we have no power to know either his nature or his existence. And because this is the case, it is impossible for anyone to solve this problem. Who would dare? Not we mere mortals, who are so disproportionate to him.

Who, then, will complain about Christians when they are unable to rationally account for their beliefs, since in truth no such account can be rendered? In proclaiming the faith to the world, they say it is a folly, a stultitiam (1 Cor. 1:18), and then you complain that they did not prove it! If they had managed to prove it, they wouldn't be keeping their word, because it's by lacking a proof that they show their sense.

"Yes," you say, "but even though this might excuse those who offer this religion as it is, and might free them from the blame of offering it without a rational argument, it doesn't excuse those who accept it."

Let's examine that point, then: let's say that God does or does not exist. Which side should we choose? Reason is powerless before such an issue. There is an infinite abyss separating us. At the far end of this infinite universe, a coin is tossed—which will turn up, heads or tails? What will you wager? Relying merely on reason, you can't decide. You can't rationally bet either way, for you can't defend either choice.

Thus, don't call people who have made a choice fools, for you know nothing about it.

"No, but I'll blame them," you say, "for making any choice at all, because, even if one man picks heads and the other man picks tails, both are equally at fault, for the right thing is not to bet at all."

But I say it's necessary to bet. You cannot avoid it, for you are already launched on the waters. This being the case, which one will you take? How will you decide? Come now, since you must choose, let's consider which one is of less importance to you. You have two things to lose—the true and the good, and two things to stake—your reason and your will, your knowledge and your bliss, and your nature has two things to shun—error and misery. Since you absolutely must choose [by living, you cannot avoid it], your intelligence will not be offended by one choice any more than by the other. That's one point settled.

But your bliss? Let's play the game and weigh the consequences of playing if you take heads—that is, that God exists. Now, let's evaluate these two cases: if you win, you win everything, and if you lose, you lose nothing; so, consider that you needn't hesitate. It's a wonderful situation!

"Yes, I must wager," you say, "but I don't want to bet too much."

> *Let's see now, since the chance of winning and the chance of losing are even, then perhaps let's say that you would win two lives for one, you could still bet, but if there were three to be won, you have to wager (since life forces you to play) and you'd be a fool, being forced to wager, not to risk your one life for three lives when the chances of losing and winning are even.*
>
> *But there is an eternity of life and happiness at stake. You have one chance of winning, and a finite number of chances of losing and what you are risking is itself finite, but what you could win is infinite. The choice is clear: there can be no excuses for timidity when an infinity of life is to be won in a game where there is a finite number of chances to lose as opposed to a single chance of winning. No cowardice, now—you must give all! And so, in a game like this, where you have to play, you would be irrational to clutch at your life rather than risk it.*
>
> *Therefore, this argument carries infinite weight, because the possibility of gaining an infinity of goodness is equal to the possibility of the loss of nothingness.*
>
> *That is conclusive and, if men are capable of truth, this statement is true!*
>
> *"I admit it, I confess, but isn't there any way of 'peeking at the cards'?" Yes, the scriptures and the rest, etc. "Yes, but my hands are tied in my lips are silenced, I'm being forced to bet and I'm not at liberty, they won't let me go, and my nature is such that I haven't the power to believe. What, then, would you have me to do?"*
>
> *It's true, but at least wake up to the fact that your inability to believe comes from your passions, since reason induces you to believe but you still can't. So, don't strive to persuade yourself by counting up proofs of God's existence; strive to diminish your passions. You want to find faith, but you don't know the way. You want to cure yourself of your unbelief and you're asking for remedies; learn from those who were once tied up like you and are now throwing the dice. They are people who know the path you'd like to follow; they are people cured of a disease from which you'd like to be cured—follow the way they started on.*
>
> *They acted as if they believed—they took holy water, they had Masses said, and the like. That will make you believe quite naturally, and will make you more pliable to the faith.*
>
> *"But that's what I fear." Why? What do you have to lose?*
>
> *How will you be harmed by choosing this path? You will be faithful, honest, humble, and grateful; you will be full of good works, and will become a true, good friend to those who know you. What will you lose? Noxious pleasures, vainglory, and riotous times, but these losses will be easily supplanted by other, greater joys.*

End of this argument.

75. The French version can be found in Pascal, *Pensées et provinciales choisies*. This is my own English translation, not that of Stanley Applebaum, who provides the English translation in this dual language edition.

76. By the way, this same argument is currently being used against Evangelical Christians because they believe in hell.

77. Jacqueline Pascal, "Letter to Soeur Angélique on the Crisis of the Signature," in *A Rule for Children and Other Writings*, 147–52.

78. Cailliet and Blankenagel, *Great Shorter Works of Pascal*, 220.

79. O'Connell, *Blaise Pascal*, 189.

80. O'Connell, *Blaise Pascal*, 192.

81. Apuleius, *The Golden Ass*, 4.31.

82. Kavanagh, *Dice, Cards, Wheels*, 9.

83. *Weird* is one of the great words, having as one of its roots "spell casting."

FURTHER READING

Armand Jean du Plessis, Duke of Richelieu and Cardinal of the Roman Catholic Church. *The Political Testament of Cardinal Richelieu*. Translated by Henry Bertram Hill. Madison: Univ. of Wisconsin Press, 1961.

Armour, Leslie. *"Infini Rien": Pascal's Wager and the Human Paradox*. Monograph published for the *Journal of the History of Philosophy*. Carbondale and Edwardsville: Southern Illinois Univ. Press, 1993.

Bennett, Deborah J. *Randomness*. Cambridge, MA: Harvard Univ. Press, 1998.

Bernstein, Peter L. *Against the Gods: The Remarkable Story of Risk*. New York: John Wiley & Sons, 1996.

Burger, A. J. *The Ethics of Belief: Essays by William Kingdon Clifford, William James, A. J. Burger*. Roseville, CA: Dry Bones Press, 1997, 2001.

Cole, John R. *Pascal: The Man and His Two Loves*. New York: New York Univ. Press, 1995.

Davies, Paul. *God and the New Physics*. New York: Simon and Schuster / Touchstone, 1983.

Davis, Philip and Reuben Hersh. *Descartes' Dream: The World According to Mathematics*. Boston: Houghton Mifflin, 1986.

Devlin, Keith. *Goodbye, Descartes: The End of Logic and the Search for a New Cosmology of the Mind*. New York: John Wiley and Sons, 1997.

Doyle, William. *Jansenism: Catholic Resistance to Authority from the Reformation to the French Revolution*. New York: St. Martin's Press, 2000.

Franklin, James. *The Science of Conjecture: Evidence and Probability Before Pascal*. Baltimore: Johns Hopkins Univ. Press, 2001.

Gray, Peter. *The Enlightenment: The Rise of Modern Paganism*. New York: W. W. Norton & Company, 1966.

Griffin, Susan. *The Book of the Courtesans: A Catalogue of Their Virtues*. New York: Broadway Books / Random House, 2001.

Groothius, Douglas. *On Pascal*. Stamford, CT: Wadsworth Division of Thomson Learning Center, 2003.

Hammond, Nicholas, ed. *The Cambridge Companion to Pascal*. Cambridge: Cambridge Univ. Press, 2003.

Heisenberg, Werner, *Physics and Philosophy: The Revolution in Modern Science*. Amherst, NY: Prometheus Books, 1999.

Houston, James M., ed. *The Mind on Fire: A Faith for the Skeptical and Indifferent, from the Writings of Blaise Pascal*. Vancouver, BC: Regent College Publishing, 1989.

Jordan, Jeff, ed. *Gambling on God: Essays on Pascal's Wager*. Lanham, MD: Rowman & Littlefield, 1994.

Kavanagh, Thomas M. *Enlightenment and the Shadows of Chance: The Novel and the Culture of Gambling in Eighteenth-Century France*. Baltimore: Johns Hopkins Univ. Press, 1993.

Levi, Anthony. *Cardinal Richelieu and the Making of France*. New York: Carroll & Graf, 2001.

———. *Louis XIV*. New York: Carroll & Graf, 2004.

McKee, Elsie Anne, ed. and trans. *John Calvin: Writings on Pastoral Piety*. Mahwah, NJ: Paulist Press, 2001.

Montaigne, Michel de. *Essays*. Translated by J. M. Cohen. London: Penguin Books, 1958.

O'Connell, Marvin R. *Blaise Pascal: Reasons of the Heart*. Library of Religious Biography. Grand Rapids, MI: William B. Eerdmans, 1997.

Pascal, Blaise. *Pensées*. Translated by A. J. Krailsheimer. London: Penguin Books, 1966.

———. *The Provincial Letters*. Translated by Thomas M'Crie. Eugene, OR: Wipf and Stock, 1997.

Popkin, Richard H., ed. *Pascal: Selections*. New York: Macmillan, 1989.

Radner, Ephraim. *Spirit and Nature: The Saint-Medard Miracles and 18th-Century Jansenism*. New York: Crossroad, 2002.

Shea, William R. *Designing Experiments and Games of Chance: The Unconventional Science of Blaise Pascal*. Canton, MA: Science History Publications, 2003.

Weaver, Warren. *Lady Luck: The Theory of Probability*. Garden City, NY: Anchor Books/ Doubleday & Company, 1963.

Wetsel, David. *Pascal and Disbelief: Catechesis and Conversion in the Pensées*. Washington, DC: Catholic Univ. of America Press, 1994.